OFDM Based Relay Systems for Future Wireless Communications

RIVER PUBLISHERS SERIES IN COMMUNICATIONS

Consulting Series Editors

MARINA RUGGIERI
University of Roma "Tor Vergata"
Italy

HOMAYOUN NIKOOKAR
Delft University of Technology
The Netherlands

This series focuses on communications science and technology. This includes the theory and use of systems involving all terminals, computers, and information processors; wired and wireless networks; and network layouts, procontentsols, architectures, and implementations.

Furthermore, developments toward new market demands in systems, products, and technologies such as personal communications services, multimedia systems, enterprise networks, and optical communications systems.

- Wireless Communications
- Networks
- Security
- Antennas & Propagation
- Microwaves
- Software Defined Radio

For a list of other books in this series, see final page.

OFDM Based Relay Systems for Future Wireless Communications

Prof. Milica Pejanovic-Djurisic

Dr. Enis Kocan

Prof. Ramjee Prasad

River Publishers

Aalborg

Published, sold and distributed by:
River Publishers
PO box 1657
Algade 42
9000 Aalborg
Denmark
Tel.: +4536953197

www.riverpublishers.com

ISBN: 978-87-92329-27-1
© 2012 River Publishers

Authors Biography

Prof. **Dr. Milica Pejanovic-Djurisic** is full professor in telecommunications at the University of Montenegro, Faculty of Electrical Engineering, Podgorica, Montenegro. Mrs. Pejanovic-Djurisic graduated in 1982. at University of Montenegro with BSc degree in Electrical Engineering. She has got MSc and PhD degrees in Telecommunications at University of Belgrade. For the period of two years, prof. Pejanovic-Djurisic has also performed research in mobile communications at University of Birmingham, UK. She has been teaching at University of Montenegro basic telecommunications courses on graduate and postgraduate levels, as well as courses in mobile communications and computer communications and networks, being the author of three books and many strategic studies. She has published more than 200 scientific papers in international and domestic journals and conference proceedings. She has organized several workshops, giving tutorials and speeches at many scientific and technical conferences. Her main research interests are: wireless communications theory, wireless networks performance improvement, broadband transmission techniques, optimization of telecommunication development policy. Prof. Pejanovic-Djurisic has considerable industry and operating experiences working as industry consultant (Ericsson, Siemens) and Telecom Montenegro Chairman of the Board. She has been in charge of GSM network design and implementation in the Republic of Montenegro. Prof. Pejanovic-Djurisic has taken part in many ICT projects with domestic and international partners.

Prof. Pejanovic-Djurisic has considerable experience in the field of telecommunication regulation. Being an ITU expert, she participates in a number of missions and ITU activities related with regulation issues, development strategies and technical solutions.

Dr. Enis Kocan is with the Faculty of Electrical Engineering, University of Montenegro, Podgorica, where he works as a teaching/research assistant since 2003. He received MSc degree in electronics engineering in 2005, and in 2011 he defended PhD thesis "Solutions for performance improvement of OFDM relay systems through subcarrier permutation," all at the Faculty of Electrical Engineering in Podgorica. Results presented in this book were obtained during his PhD research, and after that, during his postdoctoral work, under the supervision of prof. Pejanovic-Djurisic. Part of PhD research Enis has conducted at the Aristotle University in Thessaloniki. Enis has published about 40 scientific papers in international and domestic journals, international and regional peer reviewed conferences. He has participated, or is still participating, in realization of several international projects, in projects of bilateral scientific-technological cooperation and in national research projects. Besides, Enis has considerable experience in working for industry, as he has been involved in many technical projects for the telecommunication operators, in creating development strategies, elaborates, etc. His area of research and expertise are digital communications over fading channels, with the particular interest in topics:

— OFDM based relay systems for the next generation wireless communication networks,
— Multicarrier systems,
— Diversity techniques,
— Synchronization issues in wireless OFDM systems.

Prof. **Dr. Ramjee Prasad** is the Director of the Center for TeleInfrastruktur (CTIF) and Professor Chair of Wireless Information Multimedia Communication at Aalborg University (AAU), Denmark. CTIF is a large multi-area research center in telecommunication infrastructure in the premises of AAU. Under his successful leadership and based on his broad and long-term vision, CTIF currently has more than 300 scientists from different parts of the world and several CTIF branches exist worldwide, namely, CTIF-Italy (September 2006, Rome), CTIF-India (December 2007, Kolkata), CTIF-Copenhagen (March 2008, Copenhagen), CTIF-Japan (October 2008, Yokosuka), and CTIF-USA (April 2011, Princeton).

He is a Fellow of the Institute of Electrical and Electronic Engineers (IEEE), USA, the Institution of Electronics and Telecommunications Engineers (IETE), India; the Institution of Engineering and Technology (IET), UK; and a member of the Netherlands Electronics and Radio Society (NERG), and the Danish Engineering Society (IDA). He is recipient of several international academic, industrial and governmental awards of which the most recent is the Ridder in the Order of Dannebrog (2010), a distinguished award by the Queen of Denmark.

Ramjee Prasad is the Founding Chairman of the Global ICT Standardisation Forum for India (GISFI: www.gisfi.org) established in 2009. GISFI has the purpose of increasing the collaboration between Indian, Japanese, European, North-American, Chinese, Korean and other worldwide standardization activities in the area of Information and Communication Technology (ICT) and related application areas. In February 2010, CTIF under the leadership of Ramjee Prasad inaugurated the International Institute for Innovations in ICT (I4CT) in Lonavala, India, followed by a cooperation agreement signed between I4CT, GISFI, Sinhgad Technical Education Society (STES), and Aalborg University (AAU).

He is also the Founding Chairman of the HERMES Partnership (www.hermes-europe.net) a network of leading independent European research centres established in 1997, of which he is now the Honorary Chair.

Ramjee Prasad has been strongly involved in European research programs, initiating large-scale European funded international co-operations, such as the Framework Programme Four (FP4)-ACTS project FRAMES (Future Radio Wideband Multiple Access Systems), which set up the Universal Multimedia Telecommunication System (UMTS) standard and the project ASAP, the FP5 Information Society (IST) projects CELLO and PRODEMIS; the FP6 IST projects MAGNET and MAGNET Beyond.

Prof. Prasad was a business delegate in the Official Business Delegation led by Her Majesty The Queen of Denmark Margarethe II to South Korea in October 2007.

Ramjee Prasad is the founding editor-in-chief of the Springer International Journal on Wireless Personal Communications. He is member of the editorial board of several other renowned international journals and is the series editor of the Artech House Universal Personal Communications Series.

Ramjee Prasad is a member of the Steering, Advisory, and Technical Program committees of many renowned annual international conferences, e.g., Wireless Personal Multimedia Communications Symposium (WPMC); Wireless VITAE, etc. He has published more than 25 books, 750 plus journals and conferences publications, more than 15 patents, a sizeable amount of graduated PhD students (over 60) and an even larger number of graduated M.Sc. students (over 200). Several of his students are today worldwide telecommunication leaders themselves.

Ramjee Prasad's professional experience dates back to 1979. He had a pioneering role in introducing the rounded concept of code division multiple access (CDMA) as a technology for wireless personal communications, which led to the concept of the universal personal communication system, combining into one a broad range of research areas, such as radio propagation aspects, basic cellular communications including macrocellular, microcellular, and picocellular systems, adaptive equalization, multiple access protocols with CDMA concepts and random access protocols in a hostile wireless environment, dynamic channel assignment, several aspects of future public land mobile telecommunication systems (FPLMTS)/IMT-2000/UMTS and broadband multimedia communications including orthogonal frequency division multiplexing (OFDM), OFDM-based asynchronous transfer mode network and multi-carrier (MC)-CDMA. His research effort has contributed to introduce the concept of wideband CDMA (WCDMA) and a comprehensive approach to the design of a WCDMA air interface that was a major step towards the adoption of the concept of UMTS for third generation mobile communication systems. Further, his research efforts provided a comprehensive material about OFDM and how to build an OFDM demonstrator. He helped establish OFDM as a candidate technology for applications such as digital audio and video broadcasting, and wireless ATM, an effort that currently can be recognized in the significance of OFDM as the technology for fourth generation communication systems.

Contents

List of Abbreviations

3GPP	Third Generation Partnership Project
ADSL	Asymmetric DSL
AF	Amplify and Forward
BER	Bit Error Rate
BPSK	Binary Phase Shift Keying
BS	Base Station
BTB SCP	Best-to-Best Subcarrier Permutation
BTW SCP	Best-to-Worst Subcarrier Permutation
BWA	Broadband Wireless Access
CDF	Cumulative Distribution Function
COFDM	Coded OFDM
CoMP	Coordinated Multi-Point Transmission
CSMA-CA	Carrier Sense Multiple Access with Collision Avoidance
D	Destination
DAB	Digital Audio Broadcasting
DF	Decode and Forward
DFT	Discrete Fourier Transformation
DMT	Discrete Multitone Modulation
DPSK	Differentially Phase Shift Keying
DSL	Digital Subscriber Line
DSSS	Direct Sequence Spread Spectrum
DVB	Digital Video Broadcasting
DVB-T	Digital Video Broadcasting — Terrestrial
eNB	eNode B
ETSI	European Telecommunications Standards Institute

FD	Frequency Domain
FDD	Frequency Division Duplex
FDM	Frequency Division Multiplexing
FEC	Forward Error Correction
FFT	Fast Fourier Transformation
FG	Fixed Gain
Flash-OFDM	Fast low-latency access with seamless handoff OFDM
F-RS	Fixed Relay Station
GI	Guard Interval
HARQ	Hybrid Automatic Repeat Request
HIPERLAN	High Performance Radio Local Area Network
i.i.d.	independent identically distributed
ICI	Intercarrier Interference
IEEE	Institute of Electrical and Electronics Engineers
IFFT	Inverse Fast Fourier Transformation
IMT-Advanced	International Mobile Telecommunications - Advanced
ISI	Intersymbol Interference
ISM	Industrial, Scientific and Medical
ITU-T	International Telecommunication Union–Standardization Sector
LTE	Long Term Evolution
LTE-Advanced	Long Term Evolution-Advanced
MAC	Medium Access Control
MBSFN	Multicast-broadcast single-frequency network
MGF	Moment Generation Function
MIMO	Multiple Input Multiple Output
MMAC	Multimedia Mobile Access Communication
m-QAM	m-ary Quadrature Apmlitude Modulation
M-RS	Mobile Relay Station
MS	Mobile Station
N-RS	Nomadic Mobile Station
OFDM	Orthogonal Frequency Division Multiplexing
OFDMA	Orthogonal Frequency Division Multiple Access
PARP	Peak-to-Average Power Ratio
PDA	Personal Digital Assistant
PDF	Probability Density Function

PLC	Power Line Communication
R-PDCCH	Physical Downlink Control Channel
RS	Relay
S	Source
SCP	Subcarrier Permutation
SNR	Signal to Noise Ratio
TD	Time Domain
TDD	Time Division Duplex
TDMA/DSA	Time Division Multiple Access with Dynamic Slot Assignment
VDSL	Very high-bit-rate DSL
VG	Variable Gain
VLSI	Very Large Scale Integration
WiMAX	Worldwide Interoperability for Microwave Access
WLAN	Wireless Local Area Networks
WWAN	Wireless Wide Area Networks

1

Introduction

In the last decade, all segments of communication industry are highly characterized by intensive development of wireless communication systems. Thus, both WWAN (*Wireless Wide Area Networks*) and WLAN (*Wireless Local Area Networks*) networks are having a significant impact on the overall socio-economic conditions, becoming indispensable in all kind of every day activities. With the continuous demand for new web based services and multimedia applications, a significant focus is on their further development, so that the required high data rates and sufficient system capacity will be provided. At the same time, the expected traffic increase over future wireless networks is due to the explosive penetration of new smart user terminals (smartphones, tablets). Their improved processing and display characteristics represent a base for successful implementation of the well known "anywhere, anytime", communication paradigm enabling high degree of mobility in accessing new broadband services and applications.

It is well known that wireless signal transmission imposes serious challenges in fulfilling demands for high data rates and sufficient quality of service, when providing reliable communications. It is due to the complex nature of wireless radio channels whose characteristics are influenced by various types of noise, multipath effects, propagation losses, interference and other impediments, which are further on susceptible to constant change due to user mobility. Thus, in giving the adequate technical solutions for future broadband wireless networks, all relevant characteristics of this specific transmission medium have to be taken into account. One of the first steps in this direction is certainly related with the definition of optimal transmission techniques at the physical layer of wireless networks.

Over the last years, orthogonal frequency multiplexing (OFDM — Orthogonal Frequency Division Multiplexing) has been imposed as a technique which enables high data rates in severe transmission conditions encountered in wireless radio channels. It is one of those ideas that had been developing for quite a long time and it became a reality with growing demands for multimedia applications and services. OFDM modulation and transmission scheme is considered attractive for the implementation in broadband wireless networks since it mitigates the effects of multipath propagation, even in the frequency-selective fading environment. At the same time, its important advantage is in the fact that it efficiently uses limited frequency spectrum due to orthogonal subcarriers that enable spectral overlapping without interfering. With its parallel data transmission scheme, OFDM reduces various types of interference, making the use of complex equalizers unnecessary.

Due to its good characteristics, OFDM has already been implemented in a number of standardized wireless communication systems, like: DAB (Digital Audio Broadcasting), DVB (Digital Video Broadcasting), WLAN networks (IEEE 802.11a/g/n). On the other side, its extension to multiple access scheme — OFDMA, which supports multiple users by providing each of them with a fraction of the available number of carriers, has found its place in WiMAX (Worldwide Interoperability for Microwave Access) systems.

Despite the advantages of OFDM systems, it has been shown that further enhancements regarding quality and capacity of broadband communications over wireless channels are still necessary [1]. In different evolution steps towards next generation wireless communications over higher bandwidths, the increase in carrier frequencies is usually proposed. However, the available frequency spectrum is limited, with the bandwidth being a scarce resource, and it is essential to utilize it as efficient as possible. Thus, increasing spectrum efficiency becomes more and more important when upcoming technological solutions for wireless systems performance improvements are considered. At the same time, it is well known that the transmission at higher frequencies means reduction in the network coverage range due to increased propagation losses. Moreover, fulfilling the demands related with the high throughput and link reliability at the edges of coverage zones becomes specially challenging.

In that sense, it has been shown that OFDM, even in combination with the advanced techniques like multiple transmit-receive antennas, signal processing or error detection and correction codes, cannot provide sufficient capacity and the required service quality for the users located far from the transmitters (base station, access point) or near the limits of coverage areas. That is why research efforts have been directed towards new solutions and techniques that would support high data rates and higher capacities of future wireless systems, with the better coverage and link reliability at the same time.

Recently, it has been shown that cooperative communication concepts can solve many of these issues faced by future broadband WWAN and WLAN networks. In the general context, the term cooperation refers to the action of working together towards the same purpose. Here, it is a novel communication concept which means fundamental shift from traditional point-to-point communication. It is based on resource sharing and coordination among units of wireless network, which enable significant performance improvements in terms of coverage, data rates, capacity, link reliability and spectral efficiency. Cooperative communications actually use the broadcast nature of the wireless channel and allow interaction among units of wireless networks to jointly transmit information. Thus, cooperative diversity is formed leveraging the spatial diversity available between the distributed units. Its diversity gain is achieved when a number of network units collaborate and share their antennas to form a virtual multiple-input-multiple-out (MIMO) system. In such a scenario one or more intermediate nodes (terminals) intervene in the communication between a transmitter and a receiver. Thus, each terminal is supposed to transmit its own messages as well as to assist as a cooperative agent for the transmission of messages originating from other terminals. In that manner, a possibility is created for the message transmission to be realized over a better path, or the original source-destination link is kept in use but its quality is strengthened thanks to diversity provided by the cooperators. This directly leads towards resource sharing and link quality enhancement, while providing options for improving energy efficiency of wireless networks.

Cooperative communication concept can be implemented in both infrastructure based wireless networks (cellular systems, WLAN, WMAN) and infrastructure-less networks (ad hoc networks, wireless sensor networks). Depending on the nature of cooperative nodes, it can appear in different forms. Thus, the above described cooperative paradigm, where additional users (terminals) contribute to assist the source-destination communication, is usually denoted as user cooperation. Its simplest and oldest form is multi-hopping, which is actually a chain of point-to-point links, where third party users act as relays between the transmitter and the receiver. Thus, by relaying messages for each other, users enable propagation of redundant signals over multiple paths in the network, allowing the ultimate receiver at the destination to average wireless channel variations resulting from different channel impairments. Such approach is particularly interesting for applications where the use of multi-antenna arrays is problematic due to space and power limitations of terminals [1]. On the other side, cooperative communication system can also be applied to form an infrastructure network where fixed network elements (base stations, access points) are used to form a relaying network in order to improve throughput and/or coverage area.

In its elementary configuration, cooperative communication system comprises of a source station (S), a destination station (D) and a relaying station (RS). However, depending on the benefits that are expected from cooperative systems, multiple relays can be implemented in different multihop topologies. When a dual hop system with three stations (terminals) is considered, it has become very interesting due to its low implementation complexity and excellent energy saving capability [2]. Its operation is characterized with a cooperation protocol built upon two-phase transmission scheme. In the first phase, S broadcasts to the destination and to the relay terminal. In the second phase, RS performs one of the relaying techniques employed in cooperative communication systems. Those techniques can be classified as [3]:

- Transparent relaying techniques that neither modify information content of the message nor its waveform, allowing just the simple power scaling and/or phase rotation.
- Regenerative (nontransparent) relaying techniques that modify the information content or the waveform of the transmitted message.

Transparent relaying techniques include: Amplify-and-Forward relaying (AF), Linear-Process-and-Forward as well as Non-Linear-Process-and-Forward type of relaying. On the other side, some of the regenerative techniques are: Decode-and-Forward (DF), Estimate-and-Forward and Compress-and-Forward. In AF relaying, RS forwards a received signal towards the destination, after it is amplified either with fixed gain (FG) or variable gain (VG). In contrast with this approach, DF relaying means that the relay station first fully decodes a received signal, re-encodes it and then retransmits it towards the destination. Following the manner of their implementation, both AF and DF relaying techniques are usually referred to as fixed relaying cooperative protocols. They are characterized with the fact that their relay stations always forward the message received from the source. There is an alternative to fixed relaying known as selective relaying which is adapted to the conditions of the channel. In this type of relaying, a communication system reverts to non-cooperation mode each time when the chosen parameters of the communication link fall bellow certain thresholds. In those periods, communication pursues to be effectuated in a direct manner between the source and the destination.

Despite the recent surge of interest, it is worth mentioning that relay aided communication is not a new concept. First work on this issue appeared in 1969, when E.C. van der Meulen in his PhD thesis analyzed a communication channel with three terminals [8], whereas his first paper was published in 1971 [9]. However, it is only in the last decade that the interest for relay systems reappeared in the context of considering possibilities for its implementation in HiperLAN and wireless sensor networks [4–6]. When WWAN networks are concerned, numerous experimental and theoretical research works have emerged dealing with the idea of incorporating relay aided communication associated with the OFDM transmission technique [7, 10–57]. In June 2009 this resulted in completing of IEEE 802.16j standard, which is the first WiMAX standard that includes possibility of *multi-hop* communication using multiple relay stations [105]. After that, another WiMAX standard (IEEE 802.16m), as well as the next standard specified for wireless cellular networks (LTE-Advanced), have incorporated relay aided communication in order to enable full advantages of multiuser cooperative concept [112].

First considerations of OFDM based relay systems have taken into account OFDM as a solution at physical layer, while analyzing methods for optimal resource allocation, as well as power allocation per subcarrier. Subsequently, it has been noticed that specific OFDM advantages could be better used, providing that signal processing and forwarding at the level of subcarriers at RS, should take into account instantaneous subcarriers signal-to-noise ratios. Thus, a special attention has been given to possibilities for further performance enhancements of OFDM based relay systems through implementation of subcarrier permutation (SCP) technique in the signal processing/forwarding process performed at RS.

This idea about OFDM based relay system performance improvements through subcarrier permutation at RS has a substantial potential when its incorporation in future standards for WLAN and WWAN is concerned. That is why this book deals, not only with the concept and practical realization of these systems, but moreover with their in depth analyses giving apprehensive analytical tools for determining exact levels of achievable performance improvements. In that sense, the text can be considered as an important contribution in further definition of this promising technological approach. It is structured with the final goal to present how, depending on transmission conditions on particular links, full identification of optimal ways in achieving performance improvements in future wireless networks.

Thus, the book presents a comprehensive research results in analyzing behavior and performance of the OFDM based AF relay systems with fixed gain (FG) and variable gain (VG), as well as of the OFDM DF relay systems, in *dual-hop* relay scenario with three communication terminals, and no direct link between the source and the destination. This type of dual-hop relay system has been chosen as it is considered to be compliant with the conditions expected to be common for the future mobile cellular and WiMAX networks. Analyses are performed for Rayleigh narrowband fading statistics at S-RS and RS-D links, which means for the worst possible propagation case with no line of sight communication between terminals. Closed form analytical expressions, as well as simulation results, are presented, enabling bit error rate (BER) and capacity performances analyses for the considered OFDM relay systems implementing SCP at the RS, with the final goal of optimal technological solutions identification, from the point of their incorporation in the future WWAN and WLAN networks.

The book is organized as follows. Chapter 2 contains a general overview of relay techniques, with the particular attention on performances of dual-hop relay systems with AF FG, AF VG and DF signal processing at RS, all in the scenarios with Rayleigh fading statistics on both hops. Chapter 3 firstly describes OFDM concept and its basic features, and then OFDM relay system, giving an insight in the possibility of performance improvement through implementation of appropriate SCP scheme at RS. Also, in this Chapter is provided analytical derivation of the boundaries between the regions of small SNRs and high SNRs for different modulation schemes applied. These boundaries present SNR values where the considered OFDM AF relay systems should switch between the two analyzed SCP schemes in order to improve BER performances. In Chapter 4 a description of OFDM relay solutions in standardized wireless cellular networks is given. Physical layer and some aspects of link layer solutions, as well as implementation scenarios of OFDM relay systems in IEEE 802.16j, LTE-Advanced and IEEE 802.16m standards are included in this Chapter. Analytical derivations of BER performances for different modulation schemes applied, and for ergodic capacity of OFDM AF FG relay systems with SCP are contained in Chapter 5. The analytically obtained results are then compared with the simulation results, and analysis on the level of performance improvement compared to the system with no SCP is given. Analytical performance evaluation of the OFDM AF VG relay systems with SCP is shown in Chapter 6. The obtained results are completely verified through comparison with the simulation obtained results. Moreover, this Chapter comprises performance comparison of the OFDM AF VG and OFDM AF FG relay systems with SCP, in order to identify optimal solution based on AF relaying. Final Chapter 7 contains analytical derivation of BER and ergodic capacity performances of OFDM DF relay system with SCP. The obtained results are compared with the performance of OFDM AF relay systems with SCP, all with the goal to define the best solution for the future wireless communication systems.

2

General Overview of Relay Techniques

Concept of message transmission between two communication terminals using the third terminal as an intermediary between them (relay) has first appeared in 1970s. However, relay systems came into real focus in the last few years, following considerations of multi-carrier relay systems as a potential solution for new generation broadband wireless networks [4–7]. It can be noted that significant results in this field have been partially achieved due to previous accomplishments related with the research of single-carrier relay systems in different communication scenarios [59–78, 83–89]. Therefore, before explaining the multi-carrier relay concept, it is necessary to summarize the most important facts about single-carrier relay systems, which further on will serve as a base for in depth consideration of OFDM based relay systems. That is even more important knowing that, under the same communication conditions, BER performance of an ideally synchronized OFDM relay system is identical with the single-carrier relay system, while its average capacity per subcarrier is equal to the capacity of the single-carrier system [44].

This Chapter gives detailed explanations of relay aided communication concept, with the description of techniques applied for the message processing and/or forwarding at relay stations. A focus of attention is given to so called fixed relaying techniques: amplify-and-forward, with either fixed gain (AF FG) or variable gain (AF VG), and decode-and-forward (DF). Adequate analytical models leading toward BER and capacity values for a single-carrier relay system are presented. Analyses deal with the scenario with

OFDM Based Relay Systems for Future Wireless Communications, 9–31.
© 2012 *River Publishers. All rights reserved.*

three communication terminals where complete message transmission is performed using a relay terminal. Multipath fading with the Rayleigh statistics is assumed for both communication links (S-RS and RS-D). It is worth mentioning that results presented in this Chapter will be used as a reference later on, in analyzing system performance enhancements achieved with OFDM based relaying, since performances of a single-carrier relay system correspond to performances of an ideally synchronized OFDM based relay system.

2.1 Relay Based Communications

Well known characteristics of propagation over wireless communication channel (link), including the fact that signal attenuation becomes more severe with increase of distance (d) between a source and a destination, impose fundamental limitations when the transmission coverage range of wireless networks is concerned. Actually, power of a signal transmitted over wireless channels exponentially decreases when d increases, i.e., if a power emitted by transmitter is P_t, then the level of a power received at the distance d is: P_t/d^e. In this relation, the exponent e usually has values between 2 and 6, for the majority of wireless channels [59]. An interesting solution for overcoming this limitation of coverage range has appeared a couple of decades ago and it has been based on a relatively simple idea of replacing a long point-to-point communication link with a certain number of shorter links, introducing intermediary nodes between terminal transmitting a message (source) and terminal receiving a message (destination). Thus, the concept of relay aided communication has been introduced with those intermediary nodes having the role of relay stations (RS). In its simplest form, a relay based system has just one relay station and the entire communication process between a source (S) and a destination (D) is performed over RS. This represents *dual-hop* relay system with three communication terminals (Figure 2.1), where S-RS link is denoted as the first hop and RS-D link as the second hop. In the given configuration, RS receives a signal from the source, performs its adequate processing and after that transmits it toward the destination.

In order to achieve full advantages of the relay implementation, it is necessary to obtain that the communication channel between the source and RS is orthogonal with the communication channel between RS and the destination. The required orthogonality can be realized in the frequency domain, in the

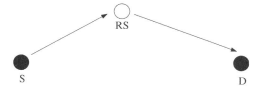

Fig. 2.1 Dual-hop relay system.

time domain, or using signals which are orthogonal in space-time constellation. In the case of S-RS and RS-D channels being orthogonal in the frequency domain, communications on both links (hops) are performed simultaneously using different frequency bands. On the other hand, orthogonality of S-RS and RS-D channels in the time domain assumes that RS operates in a half-duplex mode, and both communication processes are performed in the same frequency band, but in two different time intervals. Thus, in the first interval a signal is transmitted from S to RS, while RS forwards a signal toward D in the second time interval. Realization of relaying system which could enable operation of RS in a full duplex mode is much more complex to realize in practice, due to a very high level of interference between the signal received at RS and the one transmitted from RS. Namely, the power of the signal received at RS is usually 100–150 dB bellow the power of the transmitted signal, meaning that even the smallest error in interference cancellation would cause significant problems [61]. It is expected that further advancements in analog signal processing will enable full-duplex operation of RS, following the early works that analyzed those systems using information theory tools [9, 60]. In the meantime, research works concentrate mainly on relay systems with relay stations working in half-duplex mode [61–70].

As it has been already stated, a relay system with three communication terminals represents the simplest example of relay aided communication network and it could be considered as a particular case of a *multi-hop* relay system. It is clear that an extended wireless link, covering greater distances between S and D, cannot always be successfully realized including just one RS. If n denotes the number of relay stations participating in communication between S and D, then the multi-hop relay system is characterized with the S-D communication link being divided into $n + 1$ links (hops). There, each relay station communicates with the two neighboring terminals, as it is illustrated

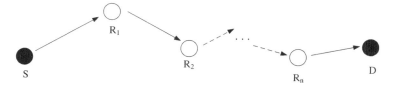

Fig. 2.2 Multi-hop relay system.

in Figure 2.2. In this case it is also necessary to maintain the orthogonality between the receiving and the transmitting channel for each relay station.

The above mentioned dual-hop and multi-hop relay systems are basically introduced in order to better cope with the effects of severe propagation losses present in wireless communications over longer distances. At the same time, their implementation contributes toward overall capacity improvements of wireless systems, enabling extension of their coverage range by maintaining the message transmission in the areas where it would not be possible without relay stations. These types of relay systems represent the first ones considered for various performance improvements of broadband wireless networks. However, relaying concept could also be implemented in a different way, providing the possibility of transmitting multiple signal replicas toward a certain destination terminal. In such circumstances standard spatial diversity is actually formed either by keeping the alternative direct S-D link, or by performing parallel relaying with a number of relay nodes each transmitting a same message received from a source. In the simplest form of three communication terminals, that would assume additional direct communication link between S and D, as illustrated in Figure 2.3.

When this scenario with three communication terminals is considered, assuming that the orthogonality between S-RS and RS-D links is achieved in the time domain, i.e., that RS operates in half-duplex mode, it is possible

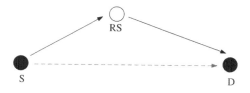

Fig. 2.3 Dual-hop relay system with diversity.

Table 2.1. Models of cooperative diversity in dual-hop relay system.

Phase 1	Phase 2	Description
S → (RS, D)	(S, RS) → D	General form of a relay system
S → RS	(S, RS) → D	D ignores signal from S in the first phase
S → (RS, D)	RS → D	S does not transmit in the second phase

to identify different models for the realization of diversity transmission. Actually, as the communication process between S and D is divided in two time intervals, or two phases depending on terminals which participate in a particular phase, three models of this cooperative diversity could be recognized (Table 2.1).

Among the presented models, the first one represents a general case that has been considered in initial analyses focused on possibilities to achieve diversity gain with relay systems consisting of single antenna terminals. It is assumed that in the first phase S broadcasts a signal toward RS and D, while in the second phase both RS and S transmit a signal to D. The other two presented models are actually particular cases of the first model and they have been introduced in order to enable less complex mathematical modeling and application of conventional space-time codes [61, 67]. Thus, in the second scenario S first transmits a signal to RS which is ignored by D, while in the second phase both S and RS transmit signals to D. The last presented model is more energy efficient since S is active only in the first phase transmitting signals to RS and D, with only RS transmitting signal to D in the second phase.

Following growing interest for MIMO (*Multiple Input Multiple Output*) systems in wireless communications, additional focus has been directed toward relaying after presenting the idea of creating virtual MIMO system using single antenna relay terminals [7]. MIMO systems, already incorporated in different wireless network standards, offer significant performance improvements of wireless systems characterized with the communication channel exposed to fading and other known impairments. However, a practical implementation of this concept might be a problem in certain conditions due to limitations related with placing multiple antennas on a single terminal. That is why virtual, or distributed, MIMO system has emerged as an interesting solution for obtaining benefits of MIMO concept in a scenario with single antenna terminals (Figure 2.4).

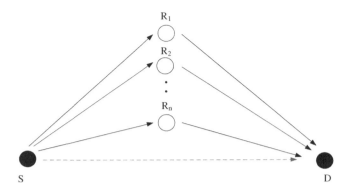

Fig. 2.4   Virtual MIMO system.

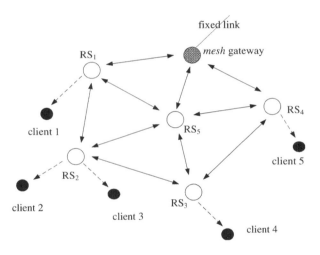

Fig. 2.5   Mesh network.

Another option for incorporating relay systems in wireless environments can be created with wireless *mesh* networks, which include mesh clients, mesh nodes and gateways (Figure 2.5). In this configuration mesh nodes actually represent relay stations that can communicate with all neighboring terminals (nodes). Thus, the existence of such redundant communication links makes mesh networks highly reliable. If a failure of a link (or a node) appears, then communication among terminals is possible over other nodes.

When expected advantages of relay systems deployment in a wireless scenario are considered, it has been shown that they depend on the first place of the level of SNR (*Signal-to-Noise Ratio*) achieved at particular communication links [66]. Apart from that, another factor which has a considerable role in dealing with deleterious effects of small-scale and large-scale wireless channel impairments is related with the type of relaying technique applied at RS.

2.1.1 Relaying techniques

With regard to algorithms of signal processing and forwarding applied at relay stations, relaying techniques can be classified as:

- Transparent relaying techniques that perform simple power scaling and/or phase rotation, i.e., linear transformation of a signal received at RS,
- Regenerative (nontransparent) relaying techniques that include modifications of a signal waveform or/and its information content, i.e., non-linear transformation of a signal received at RS.

Among the transparent techniques, Amplify-and-Forward relaying (AF) attracts most of the attention in relay systems considerations. In AF relaying, RS receives a signal from a source, amplifies it either with fixed gain (FG) or variable gain (VG), and then forwards it toward a destination. In contrast with this approach, Decode-and-Forward (DF) is typical example of regenerative relaying technique. Thus, a relay station with DF first fully decodes a received signal, re-encodes it and then retransmits it toward a destination.

Performances of the mentioned AF and DF relaying techniques highly depend on signal-to-noise ratio (SNR) of particular communication links [66], what can be considered as a limitation factor for identification of a generally optimal relaying technique. That is why a choice of an optimal relaying technique could be done only for a well defined specific communication scenario. When comparing the two relaying techniques AF and DF, considered as the most used ones, it can be noticed that AF is characterized with the simpler realization and less delay introduced at relay stations. On the other side, it has a significant disadvantage in the fact that amplifying a signal it amplifies a present noise as well, what is not the case in regenerative relaying. DF relaying has specific advantage as it allows completely separated optimizations of S-RS and RS-D links. This is due to the fact that the process of re-encoding at RS could be done with a code which is the most adequate for RS-D link no matter what was a code used for signal transmission over S-RS link. The signal processing algorithm applied in DF relaying has an important set back since it introduces a possibility of error propagation and accumulation. Namely, if RS makes a detection error in the process of a received signal decoding, an erroneous symbol will be further forwarded toward destination, causing lower performance of the overall communication system.

Taking into account the above given benefits and drawbacks of AF and DF relaying techniques, new proposals for their further improvements have emerged. Thus, considering the previously described AF and DF relaying protocols as fixed relaying techniques, selective relaying approach has been proposed as well as hybrid relaying, incremental relaying, coded cooperative relaying [7, 40, 49, 61].

Main change introduced in selective relaying is in its adaptation to current conditions of the communication channels, enabling switching from the relay aided communication to a direct link communication between a source and a destination. Thus, for the relay station with selective AF relaying technique, RS first performs an estimation of S-RS communication channel and then signal power scaling is applied as long as the power of the signal received at RS is above a certain predetermined threshold. If the signal power falls below a threshold, there is no signal forwarding toward destination, meaning that the relaying in this particular system is suspended. For the relay station with selective DF relaying technique, RS first decodes the received signal. Then, the presence of errors is checked, using redundant symbols sent by the source together with information. If the received signal is decoded with no errors, it is re-encoded and forwarded to the destination. If not, RS does not process the signal further and the communication system is back to its point-to-point mode of operation.

Hybrid relay system assumes that relay stations are equipped to perform both AF and DF relaying techniques. Which of the two signal processing schemes will be applied depends on the S-RS channel estimation and the predetermined level of SNR at the S-RS link. Further on, the level of this SNR threshold depends on the type of modulation scheme applied, as well as on the selected performance of interest, i.e., whether the goal is to minimize BER or to maximize achievable capacity.

Incremental relay technique is proposed with a goal to increase spectral efficiency of fixed relaying protocols. Namely, using a feedback from the destination terminal, RS forwards the signal only when it is necessary. Thus, terminal D sends a simple, usually one bit, information which indicates whether or not a direct message transmission between the source and the destination was successful. If the level of SNR of the S-D link is sufficiently high, the feedback information will most probably prove that the direct communication is performed successfully. In the opposite situation, the feedback information demands RS to join the process of message transmission in a way that it retransmits the signal to the destination terminal. Terminal D might then combine those two transmitted messages, introducing spatial diversity in the communication and thus increasing its reliability. In incremental relaying, when asked to participate in a message transmission, RS usually exploits AF relaying technique [61].

Another relaying technique that incorporates DF signal processing, but has its own specific features is known as coded cooperation [7]. In this approach, it is assumed that exist communication terminals which are able to, while keeping its own communication, gives a part of its communication resources to other pairs of terminals which are in direct communication. The users split their information to be sent in data blocks, each joined with cyclic redundancy check (CRC) code. Let as assume that in communication process between transmitter TR1 and receiver RC1, another transmitter TR2 gives its resources to this communication pair. At the same time, TR1 is a partner that shares its resources in communication between TR2 and receiver RC2. The transmission time of each transmitter terminal is divided into two time slots. The first one, T1, is used for sending its own data block, while a content of the second time slot T2 depends on whether or not message data block of a partner terminal is correctly decoded. Thus, if a message transmitted in T1 by a partner is correctly decoded, then this message is transmitted in time slot T2. Otherwise, in the second time slot the transmitter resends its own message from T1. The specific characteristic of this relaying technique is in the fact that a direct communication link between two terminals is kept, while neighboring terminal join the cooperative scheme in order to increase communication reliability. At the same time, the retransmission is organized with no feedback information exchange between terminals, so that the encoding process becomes the most important in controlling the communication. Since there is no feedback information, terminals are completely independent in the second time slot with no information whether the message sent in the first time slot is correctly decoded by a terminal which is a partner in cooperation. That is why four different cooperation scenarios are possible with this relaying technique:

- Both transmitter terminals participate in cooperation,
- None of the terminals participate in cooperation,
- Only the first terminal participates in cooperation (for example TR1),
- Only the second terminal participates in cooperation (for example TR2).

The implemented scenario highly defines overall performances of this coded cooperative relaying technique. Apart from that, performances are also affected by the gain introduced through coding scheme applied at terminals joining the communication as relays. Usually, an automatic repeat request (ARQ) approach is introduced and block codes, convolution codes or their combination are used.

From the explanations given above, it is clear that AF and DF relaying techniques can be considered as bases for all the other signal processing and forwarding techniques used in relay aided communications. That is why a detailed description of their behavior and performances is necessary in order to understand effects of their implementation in wireless communication networks.

2.2   Amplify and Forward Relay Technique

As it has already been mentioned, Amplify-and-Forward (AF) relay technique represents one of the two basic methods used for processing a signal received at the relay station. Here, RS receives the signal from the source and after its power scaling, retransmits it to the destination. Depending on the way the signal scaling, i.e., its amplification, is done, the following types of AF systems can be recognized:

- AF with fixed gain (FG) and
- AF with variable gain (VG).

In AF FG relaying, RS amplifies the received signal always with the same level, no matter the actual conditions on the S-RS link. On the other hand, in the AF VG system the relay station permanently estimates the S-RS link and, depending on the channel state information, determines the level of signal scaling applied.

For the elementary configuration of a dual-hop relay system with three communication terminals, shown in Figure 2.1, the signal received at RS is given as:

$$y_R(t) = x(t)h_1(t) + n_1(t), \tag{2.1}$$

with $x(t)$ being a data symbol emitted by the source at the time instant t, $h_1(t)$ is the fading amplitude of the S-RS channel and $n_1(t)$ is an additive white Gaussian noise, with variance $\mathcal{N}_{01}$.

2.2.1 AF with Fixed Gain

In AF relay systems with fixed gain G, the signal received at the destination can be represented with:

$$
\begin{aligned}
y_D(t) &= Gy_R(t)h_2(t) + n_2(t) \\
&= Gx(t)h_1(t)h_2(t) + Gn_1(t)h_2(t) + n_2(t), \tag{2.2}
\end{aligned}
$$

where $h_2(t)$ is the fading amplitude of the RS-D channel at the given instant of time and $n_2(t)$ is an additive white Gaussian noise with variance $\mathcal{N}_{02}$. The above relation (2.2) illustrates the following two important characteristics of AF FG systems: (1) if fixed gain G is neglected, the total fading amplitude at the time instant t introduced over the S-RS-D channel can be obtained by multiplication of fading amplitudes on S-RS and RS-D links at the same instant of time, i.e.:

$$h(t) = h_1(t) \cdot h_2(t), \tag{2.3}$$

and (2) these systems are characterized with the propagation of noise from RS to the destination. Usually, in the systems with the fixed gain, G is taken to be:

$$G = \sqrt{\frac{\epsilon_R}{\mathbf{E}\big[|y_R(t)|^2\big]}} = \sqrt{\frac{\epsilon_R}{\epsilon_S \mathbf{E}\big[|h_1(t)|^2\big] + \mathcal{N}_{01}}}. \tag{2.4}$$

In the relation (2.4), ϵ_R and ϵ_S denote energy of the symbols emitted by RS and S, respectively, and $\mathbf{E}(\cdot)$ is expectation operator. AF relay system implementing this type of gain at relay station is usually called *semi-blind* AF relay system, or AF relay system with the average power limitation. There, it is assumed that RS has information on the S-RS channel statistics, i.e., on the average fading power, which is assumed to have relatively slow variations. Therefore, there is no need for continual estimation of the S-RS channel. The applied gain G given with the relation (2.4) enables small variations in the average power of the signal emitted by RS over longer time interval, despite the fact that instantaneous changes of S-RS channel might be significant. Using the relation (2.2) for the analyzed AF relay system with fixed gain, the following expression for the instantaneous signal-to-noise ratio at D can be written:

$$\gamma_{end} = \frac{G^2 \mathbf{E}\left\{|x(t)|^2\right\} |h_1(t)|^2 |h_2(t)|^2}{\mathcal{N}_{02} + G^2|h_2(t)|^2 \mathcal{N}_{01}} = \frac{\frac{\mathbf{E}\{|x(t)|^2\}|h_1(t)|^2}{\mathcal{N}_{01}} \frac{|h_2(t)|^2}{\mathcal{N}_{02}}}{\frac{1}{G^2 \mathcal{N}_{01}} + \frac{|h_2(t)|^2}{\mathcal{N}_{02}}}. \tag{2.5}$$

After multiplying both numerator and denominator with the energy of the symbol transmitted by RS, ϵ_R, the above relation becomes:

$$\gamma_{end} = \frac{\gamma_{SR}\gamma_{RD}}{\frac{\epsilon_R}{G^2 \mathcal{N}_{01}} + \gamma_{RD}}, \tag{2.6}$$

where

$$\gamma_{SR} = \frac{\epsilon_S |h_1(t)|^2}{\mathcal{N}_{01}} \quad \text{and} \quad \gamma_{RD} = \frac{\epsilon_R |h_2(t)|^2}{\mathcal{N}_{02}} \tag{2.7}$$

denote the instantaneous signal-to-noise ratios of the S-RS and RS-D links, respectively. When fixed gain introduced at the relay station is the one defined in (2.4), then the instantaneous SNR at the system receiving end is given with:

$$\gamma_{end} = \frac{\gamma_{SR}\gamma_{RD}}{\frac{\epsilon_R}{\mathcal{N}_{01}} \frac{\mathcal{N}_{01} + \epsilon_S \mathbf{E}\{|h_1(t)|^2\}}{\epsilon_R} + \gamma_{RD}} = \frac{\gamma_{SR}\gamma_{RD}}{1 + \bar{\gamma}_{SR} + \gamma_{RD}}. \tag{2.8}$$

$\bar{\gamma}_{SR}$ represents the average SNR of the S-RS link. And while the relation (2.2) could lead toward the conclusion that an AF FG relay system has the total S-RS-D channel which is symmetrical ($h(t) = h_1(t) \cdot h_2(t)$), i.e., there is no difference between channels for uplink and downlink communication, the relation (2.8) shows that end-to-end SNR is not symmetrical in uplink and downlink.

Despite the fact that AF relay systems with fixed gain have come into research focus long after AF systems with variable gain, their performance have been analyzed in different communication scenarios, as well as for various types of communication channels, [62–71]. For dual-hop AF relay system, when S-RS and RS-D channels have Rayleigh narrowband fading statistics, the probability density function (PDF) of the end-to-end SNR is given as [62]:

$$f_{\gamma,end}(\gamma) = \frac{2}{\bar{\gamma}_{SR}}e^{-(\gamma/\bar{\gamma}_{SR})}\left[\sqrt{\frac{\rho\gamma}{\bar{\gamma}_{SR}\bar{\gamma}_{RD}}}K_1\left(2\sqrt{\frac{\rho\gamma}{\bar{\gamma}_{SR}\bar{\gamma}_{RD}}}\right)\right.$$
$$\left. + \frac{\rho}{\bar{\gamma}_{RD}}K_0\left(2\sqrt{\frac{\rho\gamma}{\bar{\gamma}_{SR}\bar{\gamma}_{RD}}}\right)\right], \tag{2.9}$$

where $K_0(\cdot)$ and $K_1(\cdot)$ represent the zero order and the first order, respectively, modified Bessel functions of the second kind, defined in the following manner:

$$K_0(x) = \int_0^\infty \cos(x\sinh t)\,dt = \int_0^\infty \frac{\cos(xt)}{\sqrt{t^2+1}}\,dt, \quad (x>0), \tag{2.10}$$

$$K_1(x) = \sec\left(\frac{1}{2}\pi\right)\int_0^\infty \cos(x\sinh t)\cosh t\,dt, \quad (x>0) \tag{2.11}$$

Coefficient ρ is equal to:

$$\rho = \frac{G^2\epsilon_R}{\mathcal{N}_{01}}. \tag{2.12}$$

The moment generating function (MGF) of the end-to-end SNR for dual-hop AF relay system with fixed gain can be derived as [62]:

$$\mathcal{M}_{\gamma,end}(s) = \frac{1}{(\bar{\gamma}_{SR}s+1)} + \frac{\rho\bar{\gamma}_{SR}se^{\frac{\rho}{\bar{\gamma}_{RD}(\bar{\gamma}_{SR}s+1)}}}{\bar{\gamma}_{RD}(\bar{\gamma}_{SR}s+1)^2}E_1\left(\frac{\rho}{\bar{\gamma}_{RD}(\bar{\gamma}_{SR}s+1)}\right). \tag{2.13}$$

In the above relation $E_1(\cdot)$ denotes exponential integral function:

$$E_1(z) = \int_z^\infty \frac{e^{-t}}{t}\,dt, \quad (|\arg(z)| < \pi). \tag{2.14}$$

Upper bound of the ergodic capacity for AF relay system with FG, normalized to unit bandwidth, in a scenario with Rayleigh fading channel on both hops, can be defined as [70]:

$$C = \frac{1}{2}\mathbf{E}(\log_2(1+\gamma_{end})) \le \frac{1}{2}\log_2(1+\mathbf{E}(\gamma_{end})), \tag{2.15}$$

where multiplication with 1/2 is introduced as a consequence of the fact that communication process is realized in two time intervals. Expectation of SNR at the system receiving end is:

$$\mathbf{E}(\gamma_{end}) = \bar{\gamma}_{SR} \exp\left(\frac{\theta_R}{2\bar{\gamma}_{RD}}\right) \left[2W_{-2,\frac{1}{2}}\left(\frac{\theta_R}{\bar{\gamma}_{RD}}\right) + \sqrt{\frac{\theta_R}{\bar{\gamma}_{RD}}} W_{-\frac{3}{2},0}\left(\frac{\theta_R}{\bar{\gamma}_{RD}}\right)\right],$$

(2.16)

with the coefficent θ_R being equal to:

$$\theta_R = \frac{\epsilon_R}{G^2 \mathcal{N}_{01}},$$

(2.17)

and $W_{k,\mu}(z)$ denoting Whittaker function, defined as:

$$W_{k,\mu}(z) = e^{-\frac{1}{2}z} z^{\frac{1}{2}+\mu} U\left(\frac{1}{2} + \mu - k, 1 + 2\mu, z\right)$$

$$\left(-\pi < \arg(z) \leq \pi, \ k = \frac{1}{2}b - a, \ \mu = \frac{1}{2}b - \frac{1}{2}\right).$$

(2.18)

In the above relation $U(a, b, z)$ is Kummer function defined in [79 Equations (13.1.3), (13.1.2)].

2.2.2 AF with Variable Gain

Using the relation (2.1) which describes the signal received at RS, the signal at the receiving end of the AF relay system with variable gain can be represented with:

$$y_D(t) = G(t)y_R(t)h_2(t) + n_2(t)$$

$$= G(t)x(t)h_1(t)h_2(t) + G(t)n_1(t)h_2(t) + n_2(t).$$

(2.19)

As it can be noticed, the gain applied at RS in this system is a function of time, having variations which follow changes of the S-RS channel in accordance with:

$$G(t) = \sqrt{\frac{\epsilon_R}{\epsilon_S |h_1(t)|^2 + \mathcal{N}_{01}}}.$$

(2.20)

With RSs with variable gain, it becomes possible to compensate deleterious effects related with the signal propagation over S-RS link, so that a relay station RS always transmits the signal with the same power. That is the reason why AF relay system with this type of gain is also known as AF system with the instantaneous power limitation. It is clear that the system with variable gain is more complex than the system with fixed gain, as it requires permanent estimation of the S-RS channel. Introducing the gain factor $G(t)$ into the relation (2.19), the fading amplitude of the whole S-RS-D channel at the given time t is obtained as:

$$h(t) = \frac{\sqrt{\epsilon_R} h_1(t) h_2(t)}{\sqrt{\epsilon_S |h_1(t)|^2 + \mathcal{N}_{01}}}. \tag{2.21}$$

Relation (2.21) shows that end-to-end characteristics of uplink and downlink channels are not identical. Following the expression for the signal received at the destination (2.19), SNR at the receiving end of the system with variable gain can be defined with:

$$\gamma_{end} = \frac{(G(t))^2 \mathbf{E}\{|x(t)|^2\} |h_1(t)|^2 |h_2(t)|^2}{\mathcal{N}_{02} + (G(t))^2 |h_2(t)|^2 \mathcal{N}_{01}} = \frac{\frac{\mathbf{E}\{|x(t)|^2\} |h_1(t)|^2}{\mathcal{N}_{01}} \frac{|h_2(t)|^2}{\mathcal{N}_{02}}}{\frac{1}{(G(t))^2 \mathcal{N}_{01}} + \frac{|h_2(t)|^2}{\mathcal{N}_{02}}}. \tag{2.22}$$

Introducing the variable gain $G(t)$ given in (2.20), the above relation becomes:

$$\gamma_{end} = \frac{\gamma_{SR} \gamma_{RD}}{1 + \gamma_{SR} + \gamma_{RD}}. \tag{2.23}$$

Relation (2.23) shows that, despite the fact that S-RS-D uplink and downlink channels in AF relay system with VG are not symmetrical, SNR at the system receiving end is symmetrical when SNRs at particular hops are considered. This leads to the conclusion that BER performances of this system will be identical for uplink and downlink communications.

Analyses of the AF VG relay system, focused on its capacity as well as on its bit error rate, have been performed for different communication scenarios including dual-hop, multi-hop or cooperative diversity configurations [66–75]. It has been shown that, even for the elementary dual-hop configuration, derivation of the closed form relation for the PDF of the received SNR, in AF VG relay system with Rayleigh narrowband fading statistics, might be very complex without certain approximations. That is why performance analyses of these systems usually assume that a relay station introduces variable gain $G(t)$ given as [72]:

$$G(t) = \sqrt{\frac{\epsilon_R}{\epsilon_S |h_1(t)|^2}}. \tag{2.24}$$

In comparison with $G(t)$ defined with (2.20), the above given expression does not take into account the noise present at RS. Therefore, for small SNRs on the S-RS channel, the signal gain introduced at RS is higher. On the other side, the applied gain (2.24) gives a relation for the received SNR which is much more suitable for further mathematical manipulations:

$$\gamma_{end} = \frac{\gamma_{SR}\gamma_{RD}}{\gamma_{SR} + \gamma_{RD}}. \tag{2.25}$$

Probability density function of the above defined γ_{end}, in the assumed dual-hop scenario with Rayleigh fading statistics on each particular channel (hop), can be determined as [71]:

$$f_{\gamma,end}(\gamma) = \frac{2\gamma}{\bar{\gamma}_{SR}\bar{\gamma}_{RD}} e^{-(\gamma/\bar{\gamma}_{SR}+\gamma/\bar{\gamma}_{RD})} \left[\frac{(\bar{\gamma}_{SR} + \bar{\gamma}_{RD})}{\sqrt{\bar{\gamma}_{SR}\bar{\gamma}_{RD}}} K_1\left(\frac{2\gamma}{\sqrt{\bar{\gamma}_{SR}\bar{\gamma}_{RD}}}\right) \right.$$
$$\left. + 2K_0\left(\frac{2\gamma}{\sqrt{\bar{\gamma}_{SR}\bar{\gamma}_{RD}}}\right) \right]. \tag{2.26}$$

Figure 2.1 presents graphs for probability density function of the received SNR in AF FG and AF VG dual-hop relay systems characterized with the Rayleigh fading distribution on its hops. Those graphs are obtained assuming a scenario with identical signal-to-noise ratios at S-RS and RS-D links, and the presented results are given for $\bar{\gamma}_{SR} = \bar{\gamma}_{RD} = 5\,\text{dB}$ and $\bar{\gamma}_{SR} = \bar{\gamma}_{RD} = 10\,\text{dB}$. Figure 2.6 shows that, for both AF relaying techniques, the highest probability is that the received SNR has a very small value, around $0\,\text{dB}$. At the same time it can be noticed that this probability is higher for AF VG systems, while AF FG systems have higher probability to attain the received SNR value which is very close to the average SNR value per hop.

When the MGF of the received SNR is considered, for the assumed communication scenario and the case when $\bar{\gamma}_{SR} = \bar{\gamma}_{RD} = \bar{\gamma}$, it is given with [72]:

$$\mathcal{M}_{\gamma,end}(s) = \left(\sqrt{\frac{\bar{\gamma}}{4}s\left(\frac{\bar{\gamma}}{4}s + 1\right)} + \text{arcsinh}\left(\sqrt{\frac{\bar{\gamma}}{4}s}\right)\right) \Big/ 2\sqrt{\frac{\bar{\gamma}}{4}s\left(\frac{\bar{\gamma}}{4}s + 1\right)}^{3/2}. \tag{2.27}$$

Using the PDF of the received SNR given with the relation (2.26), an upper bound of the average ergodic capacity, for Rayleigh fading statistics on both hops, is defined as [70]:

$$C = \frac{1}{2}\mathbf{E}(\log_2(1 + \gamma_{end})) \leq \frac{1}{2}\log_2(1 + \mathbf{E}(\gamma_{end})). \tag{2.28}$$

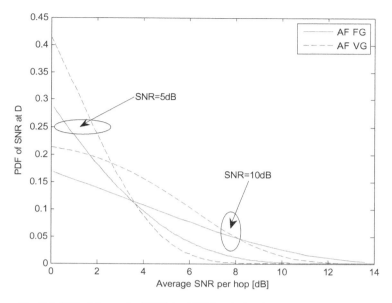

Fig. 2.6 PDF of the received SNR for AF FG and AF VG dual-hop relay systems.

SNR expectation at the receiving end, for the communication scenario considered, has the following form [70]:

$$
\mathbf{E}(\gamma_{end}) = \frac{4\sqrt{\pi}\beta_R^2}{3,3233(\epsilon_R + \beta_R)^3} \left[{}_2F_1\left(3, \frac{1}{2}; \frac{7}{2}; \frac{\epsilon_R - \beta_R}{\epsilon_R + \beta_R}\right) \right.
$$

$$
\left. + \frac{3\epsilon_R}{\epsilon_R + \beta_R} \, {}_2F_1\left(4, \frac{1}{2}; \frac{7}{2}; \frac{\epsilon_R - \beta_R}{\epsilon_R + \beta_R}\right) \right]. \tag{2.29}
$$

In the above relation, coefficients ϵ_R i β_R are equal to:

$$
\epsilon_R = \frac{1}{\bar{\gamma}_{SR}} + \frac{1}{\bar{\gamma}_{RD}} \quad \text{and} \quad \beta_R = \frac{2}{\sqrt{\bar{\gamma}_{SR}\bar{\gamma}_{RD}}}, \tag{2.30}
$$

while ${}_2F_1(\cdot, \cdot; \cdot; \cdot)$ denotes Gaussian hypergeometric function:

$$
{}_2F_1(b, a; c; z) = \frac{\Gamma(c)}{\Gamma(a)\Gamma(b)} \sum_{n=0}^{\infty} \frac{\Gamma(a + n)}{\Gamma(b + n)\Gamma(c + n)} \frac{z^n}{n}. \tag{2.31}
$$

$\Gamma(\cdot)$ is standard gamma function:

$$
\Gamma(z) = \int_0^{\infty} t^{z-1} e^{-t} dt, \quad (Re(z) > 0). \tag{2.32}
$$

2.3   Decode and Forward Relay Technique

When AF relay techniques are analyzed, the S-RS-D channel can be considered as a whole, i.e., end-to-end channel fading amplitude $h(t)$ at the given instant t can be introduced, as it has been shown for dual-hop AF FG and AF VG relay systems through relations (2.3) and (2.21), respectively. On the other side, in decode and forward technique dual-hop communication is performed over two completely separated subchannels, since RS first decodes the signal received from the source and then the signal is re-encoded at RS and transmitted to the destination. If the signal received at RS is represented with:

$$y_R(t) = x(t)h_1(t) + n_1(t), \tag{2.33}$$

then the signal received at the destination becomes:

$$y_D(t) = \hat{x}(t)h_2(t) + n_2(t), \tag{2.34}$$

where $\hat{x}(t)$ denotes an estimation of the signal $x(t)$, obtained at RS. Decoding process which is applied at RS introduces evident system performance improvements since the total noise at the destination is decreased when compared with AF relay systems. At the same time, it becomes possible to implement modulation schemes at S-RS and RS-D links which are not necessarily identical, so that optimal modulations can be applied in accordance with SNR levels at particular links. Thus, communication process has to be divided in two asymmetric time intervals, where the longer time interval is always dedicated to the communication process over the link with smaller SNR. This presents another advantage of the system with DF relaying, when compared with AF systems, and it is clear that it leads toward better BER performance. On the other side, DF signal processing at RS can be at the origin of certain drawbacks in the case of channels with severe fading. When BER is concerned, a degradation appears if there is an error in the decoding process engaged at the relay station, i.e., when the characteristics of the S-RS channel cause high error probability at RS. The problem is introduced with such erroneously decoded symbols which are then further transmitted to the destination. There is no doubt that characteristics of RS-D link can also contribute to overall BER performance degradation of DF relay system, as additional errors might be introduced in the signal decoding at terminal D. Thus, effects of error propagation from RS to D are always strongly emphasized, since they might even block D from receiving the original message if it receives the signal only from RS (i.e., there is no S-D direct communication), no matter the situation at the RS-D link. In order to reduce those negative implications of the error propagation and to improve BER performance, different encoding schemes for error detection and correction can be applied in the process of signal regeneration at DF relay stations [76].

When achievable capacity of DF dual-hop relay system is concerned, it is of the uttermost importance to notice that it is limited with the characteristics of the worse of the two links engaged in the communication process. Namely, ergodic capacity of DF dual-hop relay system can not be higher than the ergodic capacity of the link (S-RS or RS-D) which has the lower instantaneous signal-to-noise ratio, i.e.:

$$C = \frac{1}{2} \min \left\{ \log_2(1 + \gamma_{SR}), \ \log_2(1 + \gamma_{RD}) \right\}. \tag{2.35}$$

2.4 Performance of AF and DF Relay Systems

In order to pursue with analyses of OFDM relay systems, it is necessary to systematize facts and numbers related with the performance of single-carrier relay systems described in this chapter. Moreover as it is well known that BER performances of ideally time and frequency synchronized OFDM relay systems are identical with BER performance of single relay systems, while the average capacity per subcarrier of the OFDM relay system is equal with the capacity of the single-carrier relay system [44].

BER of the single-carrier AF relay system can be determined using the known MGF of SNR at the system receiving end [82]. Thus, for example if DPSK modulation is applied, bit error rate is defined with:

$$P_b = 0, 5 \mathcal{M}_{\gamma,end}(1). \tag{2.36}$$

Introducing (2.13), or (2.27) into relation (2.36), BER expressions are obtained for AF relay system with fixed gain and AF system with variable gain, respectively.

As it has been already mentioned, in DF relay systems a signal is transmitted over two cascade links and its decoding is done twice. If the transmission implies a binary signal with two possible symbol states (DPSK or BPSK), an error will appear at the final destination terminal only if an error in the signal detection is performed once (either on the first or on the second link). Thus, the overall probability of error is given as:

$$P_b = 1 - \left[(1 - P_{b1})(1 - P_{b2}) + P_{b1}P_{b2}\right] = P_{b1} + P_{b2} - 2P_{b1}P_{b2}, \quad (2.37)$$

where P_{b1} and P_{b2} are bit error rates at the first and the second link (hop), respectively. When DPSK is assumed, probability of error for each of the links is given with (2.36) and, for Rayleigh fading statistic on both links, the overall BER is obtained in the form:

$$\begin{aligned}
P_b &= \frac{1}{2}\left[\left(\frac{1}{1 + \bar{\gamma}_{SR}}\right) + \left(\frac{1}{1 + \bar{\gamma}_{RD}}\right) - \left(\frac{1}{1 + \bar{\gamma}_{SR}}\right)\left(\frac{1}{1 + \bar{\gamma}_{RD}}\right)\right] \\
&= \frac{1 + \bar{\gamma}_{SR} + \bar{\gamma}_{RD}}{2(1 + \bar{\gamma}_{SR})(1 + \bar{\gamma}_{RD})}. \quad (2.38)
\end{aligned}$$

Figure 2.7 illustrates BER performance for DPSK modulated DF and AF dual-hop relay systems operating in the assumed scenario with Rayleigh narrowband fading statistics on S-RS and RS-D links. For AF relay systems, BER values are given for both schemes, i.e., for relay stations fixed gain as well as for relay stations with variable gain. Presented results are obtained

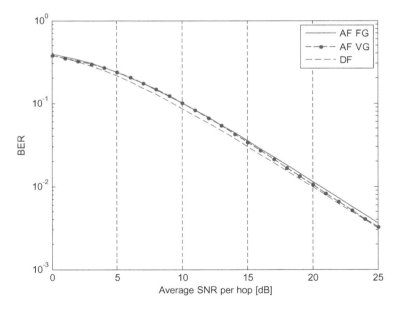

Fig. 2.7 BER performance comparison for AF and DF dual-hop relay systems.

using analytical relations (2.36) and (2.38), with the introduction of appropriate expressions for the MGF of the received SNR. It is also assumed that the average SNR of the S-RS link is equal to the average SNR of the RS-D link, i.e., $\bar{\gamma}_{SR} = \bar{\gamma}_{RD}$.

Results presented in Figure 2.7 give interesting and in a certain manner surprising results, since differences among BERs for the three systems considered are unexpectedly small, having in mind considerable differences related with the complexity of the systems and their implementations. Still, it can be seen that DF relay technique has the best BER performance. However, even when compared with AF FG system which has the worst BER performance, SNR gain does not overpass 0.5 dB for the whole range of BER values analyzed. At the same time, for higher SNRs per hop (over 24 dB), BER results for AF VG relay system are identical with the ones obtained for DF relay technique. Direct comparison of AF FG and AF VG relay systems shows that AF VG system has better BER performance for small SNRs, while performances are almost identical in the range of medium SNRs. For SNR per hop above 13 dB, AF VG relay system again outperforms AF relay system with fixed gain.

In order to gain a complete insight into benefits and trade-offs related with the choice of a particular signal processing/forwarding technique applied at relay stations, it is necessary to take into account achievable capacity as well. Figure 2.8 shows appropriate ergodic capacities of AF FG, AF VG and DF relay systems. Presented results are obtained by simulation under assumption that S-RS and RS-D links are characterized with Rayleigh narrowband fading statistic with the average SNR at the first hop being equal to the average SNR at the second hop.

It can be clearly noticed that AF VG relay system achieves the lowest ergodic capacity for the whole range of SNRs per hop, excluding very small SNRs (up to 2.5 dB) where its capacity performance is slightly better when compared with AF system with fixed gain. DF relay system has the highest ergodic capacity for SNRs per hop bellow 12.5 dB, while for higher SNRs it is AF FG system which has the best values of ergodic capacity. Following the same logic as the one applied in analyzing BER performance, surprisingly small differences among capacities of the three systems can be noticed. However, it is important to emphasize the advantage of AF FG relay systems for medium and high values of SNRs per hop, especially having in mind relatively simple physical realization of this system. Thus, for example, the presented

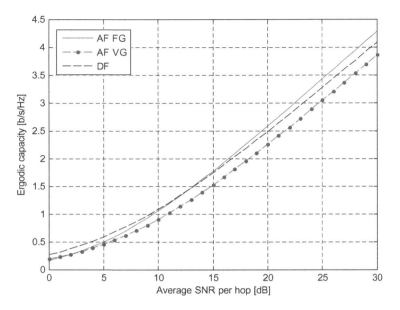

Fig. 2.8 Comparison of ergodic capacities for AF and DF dual-hop relay systems.

graphs show that for ergodic capacity being equal to 3 b/s/Hz, AF FG relay system has SNR gain of almost 1 dB in comparison with DF relay system, while its SNR gain in comparison with AF VG relay system is a bit less than 2.5 dB.

Following the description of individual relaying techniques given at the beginning of this chapter, as well as the above given comparison of theirs BER and capacity performances, it is quite clear that it is not possible to identify a technique which would be absolutely superior in terms of performances for the whole range of SNR values. However, depending on characteristics of S-RS and RS-D links and on a performance being of interest for a specific communication process, there is always a possibility to assume which of the three analyzed systems will provide the best transmission quality and reliability. Still, it is important to have in mind that, when BER and capacity are concerned, differences between the three systems are very small. Further on, taking into account the system complexity for each of the single-carrier relay systems considered in these analyses, a certain advantage of AF FG relay system can be recognized.

The interest in realization and behavior of single-carrier relay and cooperative systems, in various communication conditions, dates back to early 1970's, when the upper and lower bounds of achievable capacity for relay channels have been determined [9]. However, in recent years more and more research efforts are related with relay systems which apply *multi-carrier* techniques at the physical layer. This emerging interest is mostly focused on the technique of orthogonal frequency multiplexing (OFDM), already well known for its potentials in achieving high speed data transmission over wireless communication channels, even in scenarios with severe frequency-selective fading. Following their ability to extend the coverage of a base station, i.e., the range of a wireless local network, relay systems incorporating OFDM have already been included in IEEE 802.16j standard for WiMAX (*Worldwide Interoperability for Microwave Access*) networks, which incorporates multi-hop relay forwarding [105]. At the same time, OFDM is a part of the IEEE 802.11s WLAN standard, which is focused on WLAN mesh networks [58]. 802.16j standard deals with infrastructure relay station deployment, where relay stations are used only for message forwarding, while the IEEE 802.11s standard assumes that user terminals could give a part of their resources to another pair of communication terminals, thus performing a relaying function. Further improvements concerning multi-carrier relaying techniques have been introduced through the process of the IEEE 802.16m WiMAX standard formulation, as well as through the work related with the evolved mobile cellular networks.

Therefore, having in mind extremely dynamic activities in finding the most appropriate solutions for future wireless network standards, it is very important to understand the idea behind multi-carrier relaying techniques. These techniques based on OFDM are already seriously considered to be included in new versions of standards for WLAN, while they are already defined as a solution for the future WiMAX and mobile cellular networks, where dual-hop relaying through infrastructure relay stations has been extensively analyzed. At the moment, most of the research work is focused on the optimization of this type of communication paradigm, [49–58, 108–110], what is at the origin of motivation to give necessary explanations, and in depth analyses of their performance, in the following chapter.

3

OFDM Relay Systems

Following its potentials in supporting high speed data transmissions, even for communications over channels with severe frequency-selective fading, as well as its robustness to deleterious effects of multipath signal propagation, OFDM has been incorporated in numerous wireless communication systems: WLAN networks, WiMAX systems, LTE (*Long Term Evolution*) systems, etc. Its combination with relaying techniques has already been recognized as a solution for improving reliability and quality of service for mobile users located at the edges of the base station coverage area, which has been formulated through adoption of the IEEE 802.16j standard for mobile WiMAX multi-hop relay systems [105]. Further on, it is expected that the new generation of WiMAX systems, and mobile cellular systems, will apply OFDM based relay forwarding [49–55].

That is the reason for recent surge of research interest in OFDM based relay systems, with still a number of open issues to be addressed, related with theoretical aspects, performance improvements and practical implementation. When solutions for OFDM based relay systems performance improvements are concerned, a proposition based on introduction of subcarrier permutation at relay stations seems to be very promising. There, in the case of dual-hop OFDM relay system, the relay station performs mapping of subcarriers from S-RS link to subcarriers on RS-D link, depending on their instantaneous SNRs, enabling thus achievement of maximal capacity and/or minimal BER of the system, [20–39]. With the focus on such type of OFDM relay systems, this

OFDM Based Relay Systems for Future Wireless Communications, 33–57.

book mainly deals with the implementation of the appropriate subcarrier mapping scheme at RS and investigations of its impacts on the overall system performance improvements. Thus, the presented analyses are directed towards identification of an optimal OFDM relaying system configuration which can be then considered for incorporation in future high speed multimedia wireless communication networks.

This chapter starts with a brief description of OFDM transmission concept, summarizing its advantages and disadvantages. Then, current and future deployment scenarios for OFDM are presented, including the standardized 802.16j networks, as well as propositions for the next generation WWAN networks. An overview of up-to-date research work and results related with OFDM based relay systems is produced, with the special focus on the systems using subcarrier permutation scheme at the relay stations. At the end of this chapter, detailed analyses of the RS subcarrier permutation concept are given, including its possible impacts on the performance improvement of the OFDM based relay systems.

3.1   Basic OFDM Principles

Orthogonal frequency division multiplexing (OFDM) modulation and transmission technique has drawn considerable attention in the last decade and it has already become widely applied in new communication systems for high speed data transmission. Basic idea behind OFDM is in achieving parallel data transmission using subcarriers which are mutually orthogonal. Using serial-to-parallel convertor, an input data stream of the high data rate is divided into M parallel streams having M times lower data rates. Data in parallel branches modulate mutually orthogonal subcarriers and then the obtained modulated signals are summed. Orthogonality between subcarriers assumes that data symbols in each of the parallel branches contain integer number of carrier periods. Figure 3.1 illustrates the above described concept of the OFDM modulation.

The input data stream can be m-QAM (*m-ary Quadrature Amplitude Modulation*) or m-PSK (*m-ary Phase Shift Keying*) mapped, where m denotes the number of states of higher order modulated digital signal. If duration of symbols in parallel branches is T, and M is the number of parallel branches,

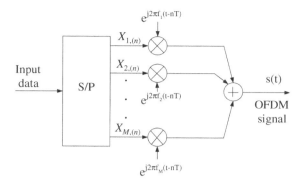

Fig. 3.1 Concept of an OFDM modulator.

then the signal $s(t)$ at the output of OFDM modulator can be presented as:

$$s(t) = \sum_{n=-\infty}^{\infty} \sum_{i=1}^{M} X_{i,(n)} e^{j2\pi f_i (t-nT)} w(t - nT) \qquad (3.1)$$

If the time domain summation is omitted, similarity of the relation (3.1) with the inverse Fourier transformation is obvious. In this relation, $X_{i,(n)}$ is the information symbol in the i-th parallel branch, in the n-th signalization interval, f_i is the subcarrier frequency in the i-th parallel branch, and $w(t)$ represents window function given with:

$$w(t) = \begin{cases} 1, & 0 < t \leq T \\ 0, & t \leq 0, \, t > T \end{cases} \qquad (3.2)$$

The subcarrier frequency, as well as subcarrier spacing (F), are defined as:

$$f_i = \frac{i - 1}{T}, \quad F = \frac{1}{T} \qquad (3.3)$$

The above relations (3.4), for subcarrier frequencies in parallel branches of OFDM modulator, assure mutual orthogonality of subcarriers. Thus, if oscillators at the transmitter and the receiver ends are completely synchronized, the appearance of interference among subcarriers (ICI — *Intercarrier Interference*) is eliminated and the OFDM signal has the spectrum illustrated in Figure 3.2.

OFDM demodulator performs the inverse operation, based on discrete Fourier transformation (DFT — *Discrete Fourier Transformation*). Following

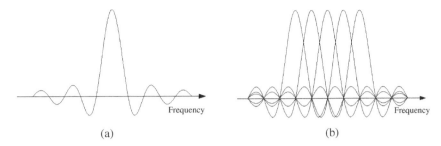

Fig. 3.2 (a) A subcarrier in frequency domain, (b) part of the OFDM signal spectrum.

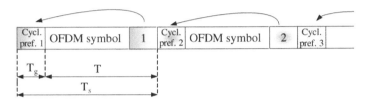

Fig. 3.3 Cyclic prefix in OFDM symbol.

the introduction of an algorithm for fast Fourier transformation (FFT — *Fast Fourier Transformation*), the signal processing time is reduced in the OFDM transmitter, as well as in the receiver, since the number of necessary operations decreased from M^2 for DFT, to $M \log_2 M$ for FFT [90].

In order to combat deleterious effects of multipath fading, and to prevent appearance of intersymbol interference (ISI — *Inter Symbol Interference*), OFDM technique implies introduction of a so called guard interval, added to each OFDM symbol. Duration of the guard interval (T_g) has to be greater than the maximal delay spread introduced by the channel. As this guard interval contains copy of the last part of an OFDM symbol, it is also known as cyclic prefix (Figure 3.3).

Besides being introduced in order to eliminate ISI, the cyclic prefix can be used for time, as well as for frequency synchronization of the OFDM transmitter and receiver. In both cases, it is based on the periodicity introduced in OFDM symbols by cyclic prefix. The size of the cyclic prefix may vary in dependence on the channel characteristics, and usually is in the range from one tenth up to one quarter of the OFDM symbol duration. It is also worth mentioning that the consequence of the guard interval introduction is

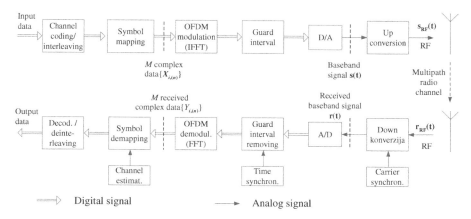

Fig. 3.4 Block diagram of an OFDM system.

certain redundancy, as this part of the OFDM symbol does not carry useful information.

3.1.1 OFDM System Structure

Figure 3.4 illustrates block diagram of an end-to-end OFDM system.

At its input, there is an optional block that performs channel coding and interleaving, and it is incorporated in order to improve the system performance. Actually, in the frequency-selective fading environment, there is a high probability that signals from one or more adjacent sub-channels could be significantly attenuated, what could cause decision errors in the receiver. Implementing some of the codes for error correction (FEC — *Forward Error Correction* codes), and interleaving, in the case when more adjacent sub-channels are exposed to the same severe fading, the probability of error in the receiver is reduced. Those OFDM systems which use codes for error correction are usually known as COFDM (*Coded* OFDM) systems.

Block for the symbol mapping performs mapping of the input data stream in the higher-order PSK modulated symbols (8-PSK, 16-PSK,...) or quadrature amplitude modulated symbols (QAM, 16-QAM, 64-QAM,...). Signal at the transmitter, on the left side from the block performing OFDM modulation, is given in the frequency domain (FD). After the IFFT (*Inverse* FFT) operation, the signal is transformed to the time domain (TD). It is assumed that the OFDM modulation block performs serial-to-parallel conversion before the

IFFT operation. $X_{i,(n)}$ ($i = 1, \ldots, M$, where M corresponds to the number of the parallel branches, i.e., to the size of IFFT transformation) is a complex data symbol in the n-th signalization interval, on the i-th parallel branch at the input of the IFFT block. In the real OFDM systems some of the subcarriers are not used, like the ones at the ends of the spectrum occupied by the OFDM signal, due to the low pass filters used for digital/analog (D/A) conversion at the transmitter and analog/digital (A/D) conversion at the receiver.

After the signal conversion to the time domain, a guard interval is added to each OFDM symbol. The size of the guard interval is chosen in such a way to be greater than the maximum value of the delay spread, in order to avoid ISI. After that, D/A conversion is performed and the signal is up converted in the desired spectrum range.

The receiver performs operations that are inversed to those conducted in the transmitter. One of the most important issues in the receiver design is in accomplishing time and frequency synchronization with the transmitter. If the receiver does not acquire the necessary level of the synchronization with the transmitter, then ISI and/or ICI will occur. For the coherent detection, equalization must be undertaken at the receiver, and it can be achieved by zero-forcing method. On the other side, for the differential modulations, the equalization is omitted, as the differential detection is performed. The differential modulations in OFDM systems can be done in the time domain (among the corresponding subcariers in consecutive OFDM symbols), or in the frequency domain (among the adjacent subcarriers of an OFDM symbol).

A particularly interesting fact about OFDM system is that it can be presented as a set of parallel Gaussian channels, if it is ideally synchronized. This means that ideally synchronized OFDM system enables to split the frequency-selective channel into the set of parallel Gaussian channels with the frequency-flat fading (Figure 3.5). The channel imposes attenuation and phase rotation on each subcarrier, what is presented through the subcarrier channel transfer functions $H_{i,(n)}$.

3.1.2   Benefits and Shortcomings of OFDM

From all the above mentioned, the main characteristics that made OFDM modulation and transmission technique a preferred solution for the

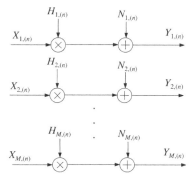

Fig. 3.5 The equivalent model of ideally synchronized OFDM system.

implementation in many different modern communication systems, can be recognized as:

* High spectral efficiency due to the subcarrier orthogonality;
* Efficiency in combating multipath fading effects, achieved through implementation of the guard interval of the size greater than the delay spread;
* Robustness against the frequency-selective fading, due to equalizations performed on the subcarrier level, based on estimates of channels for each subcarrier.
* Robustness to narrowband interference, as it affects a very small percentage of subcarriers.
* In slow time-varying channels it is possible to significantly increase the capacity, by adopting the data transmission speed on each subcarrier in accordance with the instantaneous SNR.

The concept of OFDM technique, which has brought many benefits, is also characterized with some drawbacks. First of all, this is related to the sensitivity of OFDM systems to frequency offset and phase noise, as well as to high peak-to-average power ratio (PAPR). Frequency offset violates subcarrier orthogonality thus causing ICI, which has as a consequence significant BER performance degradation. The problem of frequency offset estimation and correction has been a subject of research interest for a long time, and it still presents an actual topic, [91–94]. In general, frequency offset estimation methods can be divided in two groups:

- methods based on pilot sequences,
- methods using redundancy inherently embedded in each OFDM symbol, due to the presence of the cyclic prefix.

High peak-to-average power ratio reduces the efficiency of RF amplifiers at the output of OFDM transmitter, thus increasing the number of intermodulation products that can cause problems at the receiving end. There are different approaches for PAPR reduction, like selective mapping, cyclic coding, implementation of Golay sequences, etc., [95]. This issue is not concluded yet, as there are still ongoing research efforts aiming to find solutions for increasing the output amplifier efficiency.

3.1.3   Implementation of OFDM and Perspectives

In early 1980's a very serious problem in Digital Audio Broadcasting (DAB) was the multipath signal propagation effect. Until OFDM was realized in digital technique, this kind of service could not be used in outdoor environment for mobile users. The first DAB standard, which also represents the first standardized system based on OFDM, has been accepted in 1995. Significant experiences obtained through these systems, as well as the great success in their implementation, served as a catalyst for the adoption of standard for digital video signals broadcasting (DVB-T — *Digital Video Broadcasting — Terrestrial*, standard EN 300 744) in 1996 in Europe, [90, 100].

Soon after that OFDM has been implemented in WLAN networks. The IEEE 802.11a standard has been adopted in 1999 and it represents the first standard where OFDM is used with packetized data transmission. In accordance with the decision of the IEEE 802.11 standardization group, ETSI (*European Telecommunications Standards Institute*) has implemented OFDM in the HIPERLAN/2 (*High Performance Radio Local Area Network*) standard for WLAN networks. Similarly, in Japan the MMAC (*Multimedia Mobile Access Communication*) OFDM based standard is accepted. As the IEEE 802.11a standard, operating at 5,7 GHz, did not have the expected success, the IEEE standardization group has developed 802.11g OFDM based standard, operating at 2,4 GHz and enabling data rates up to 54 Mb/s. The last adopted

OFDM based WLAN standard is denoted as IEEE 802.11n (September 2009) and it specifies the possibility of using multiple antennas at the transmitter and at the receiver side (MIMO — *Multiple Input Multiple Output*), as well as the possibility of using 40 MHz wide channel [96].

OFDM as a multiple access technique (OFDMA — *Orthogonal Frequency Division Multiple Access*) is being used in the current IEEE 802.16 standards. These standards have been defined with the aim to support broadband wireless access (BWA), i.e., to give an alternative to cable and DSL modem connections, which are used in wired access networks. The first IEEE 802.16 standard has been adopted in 2001, but it was single carrier system. Since 2003, when the IEEE 802.16a standard was accepted, OFDM is implemented in all IEEE 802.16 standards. While the work of the IEEE 802.16 standardization group was ongoing, WiMAX (*Worldwide Interoperability for Microwave Access*) Forum was founded. The Forum included equipment manufactures whose goal was the promotion of this technology and certification of the equipment compatibility with the IEEE 802.16 standards. The first WiMAX solutions were defined in order to provide services to fixed users, and the first standard that assumed provision of services for mobile users was the IEEE 80216e standard, adopted in 2005. The current standard IEEE 802.16-2009 was adopted in May 2009, and it represents revision of the IEEE 802.16-2004 which comprises all the valid amendments [97].

Flash-OFDM (*Fast low-latency access with seamless handoff OFDM*), also known as F-OFDM, represents another practically implemented OFDM based communication technology. F-OFDM is a solution for packet-switched cellular systems operating in 450 MHz band, which was previously occupied by analog mobile systems like NMT-450 (*Nordic Mobile Telephone*) and C-Net 450 in Germany. Mobile operator Digita began with implementation of F-OFDM based cellular network in Finland in April 2007. T-Mobile in Slovakia has covered an area inhabited by 70% of the country population by F-OFDM based network, with the offered data speeds of 5,3 Mb/s on downlink and 1,8 Mb/s on uplink [98].

OFDM has also found its place in many wired communication systems. Modulation technique similar to OFDM, denoted as (DMT — *Discrete Multitone Modulation*), has been developed for stationary channels, like the digital subscriber line (DSL) channel. DMT uses DFT or FFT operation and guard interval at the transmitter, like OFDM, but it has ability of bit-loading in accordance with the each subcarrier channel state. This kind of modulation technique is implemented in ADSL (Asymmetric DSL) systems, whose standard G.992.1 was adopted in 1999, then in the systems presenting the upgrades of the ADSL, as well as in the VDSL (Very high-speed DSL) systems [90].

In June 2005, in the frame of the IEEE standardization body, the P1902 working group was founded, with the aim to develop PLC (Power Line Communication) standard which will enable high data rate (more than 100 Mb/s at the physical layer) transmission over power lines, using 100 MHz wide bandwidth [99]. Draft of the IEEE P1901 standard was published in January 2010, and it predicts one MAC layer and possibility to support two types of physical layers: one based on *wavelet*-OFDM, and another based on *windowed* FFT OFDM [99].

Through implementations in different standardized communication systems, OFDM has proved itself as a modulation technique which is very robust to different impairments and problems that may occur on wired and wireless channels, providing at the same time very high data rates. Besides, the experiences gathered through the years of practical implementations have helped to find new solutions for further OFDM performance improvements. Because of its good characteristics OFDM has already been applied at a physical layer for downlink communication in LTE (*Long Term Evolution*) systems. LTE represents a result of hard work of 3GPP (Third Generation Partnership Project) working group, which has begun in November 2004. Final specifications are defined in Release 8 document, concluded in December 2008 [100]. According to this document, LTE should support data rates of 100 Mb/s on downlink and 50 Mb/s on uplink, less than 10 ms round trip time (RTT) on the radio access part, possibility of using different bandwidths, as well as TDD (Time Division Duplex) and FDD (Frequency Division Duplex) communication. It has been shown that these requirements can be achieved using OFDM in combination with MIMO technique [101].

For the mobile cellular systems of the fourth generation (4G), OFDM is also accepted as a transmission technique [49, 107]. Namely, in October 2009, 3GPP has proposed to ITU-T its solution for the mobile communication systems of the next generation, which satisfies all the IMT-Advanced requirements. This system, denoted as LTE-Advanced, represents further improvement of the LTE system, and it achieves the spectral efficiency above 15 b/s/Hz on downlink and above 6,75 b/s/Hz on uplink, as well as peak data rates above 1 Gb/s for the slow mobile users, thus fulfilling LTE-Advanced speci-fications. LTE-Advanced systems will implement many advanced solutions, like carrier aggregation, MIMO techniques, coordinated multi-point transmis-sion and reception, and relay techniques for extension of the coverage area and enhancement of system capacity [108].

Another technology accepted for 4G systems is mobile WiMAX of the next generation (WirelessMAN-Advanced), specified through the IEEE 802.16m standard. This standard represents improvement of the first mobile WiMAX standard IEEE 802.16e-2005, and it meets all the IMT-Advanced require-ments [109]. Like the previous WiMAX systems, the IEEE 802.16m systems will implement OFDMA on the physical layer, and especially interesting option is the possibility of multi-hop relaying. The multi-hop relaying as a solution for the coverage extension has been already specified in the previous IEEE 802.16j-2009 standard.

3.2 Overview of OFDM Relay Systems

Performance of single-carrier relay systems in the case of frequency-flat fading channels have been examined in details for different communication scenar-ios [59–78, 83–89]. As OFDM has already proven its good characteristics for the data transmission over wireless frequency-selective fading channels, it came as natural to be considered as a solution for multi-carrier wideband relay systems. At first, the possibility of OFDM based relay system implementa-tion in HYPERLAN/2 wireless networks was analyzed, with the goal of the access point coverage extension, as well as for the improvement of achiev-able capacity. Soon after that, the concept of OFDM based relay systems has been generalized to all wideband wireless systems. It has been shown that, in the case of ideally time and frequency synchronization of all communica-tion nodes, BER performance of OFDM based relay systems are the same as the BER performance of single-carrier systems in the same channel condi-tions [44]. At the same time, the capacity of OFDM relay systems is higher than the capacity of single-carrier systems, due to data transmission on parallel orthogonal subcarriers.

All initial considerations of OFDM were characterized with the common approach which dealt with OFDM as a solution on physical layer that could successfully mitigate the deleterious effects of multipath fading, enabling at the same time the overall system capacity improvement. However, further benefits that can be achieved by using OFDM in relay systems have been soon recognized, especially for situations when relay station knows information about the channel state for all subcarriers on S-RS and RS-D links. Thus, OFDM as a transmission technique enables RS to achieve performance improvement through making certain decisions at the level of subcarriers, depending on characteristics of subcarrier channel transfer functions.

One of the first ideas in using this crucial possibility has been presented in [40], where a hybrid OFDM relay scheme has been proposed, with RS taking the decision whether to perform AF or DF relaying at the level of subcarriers. Half-duplex RS estimates $\gamma_{k,SR}$ for all subcarriers on the first hop, while the information on $\gamma_{k,RD}$ for the subcarriers on the second hop is obtained from terminal D. Using these information RS calculates BER values that would be achieved in the cases when DF or AF would be deployed on the given subcarrier, as well as for the case with no relaying. RS decides on the subcarrier basis to use the scheme which provides the best BER performance. The proposed system achieves significant BER performance improvement compared to classic AF and DF relaying systems, but at the price of the increased system complexity and signaling overhead. Thus, for example, when $\bar{\gamma}_{SD} = 8\,\mathrm{dB}$ and $\bar{\gamma}_{RD} = 20\,\mathrm{dB}$ this hybrid OFDM relay system provides SNR gain of about 6 dB compared to the AF relay system, and SNR gain of 7 dB compared to the DF relay system.

It is also possible to use the advantages of OFDM in relay systems comprising more relay terminals. Thus, in multi-hop relay system, decisions can be made on a subcarrier basis about the relay terminal which should be used for relaying, thus significantly decreasing outage probability. This concept can be applied even in dual-hop OFDM based relay system if more terminals which can take part in the relaying process exist.

When realization of full advantages of OFDM based relay systems is considered, a number of interesting solutions for getting the required level of system performance can be recognized. One of them deals with implementation of an appropriate power allocation algorithm per subcarrier at S-RS and RS-D links. This type of approach can be applied for both, AF and DF, relaying techniques, resulting mainly in capacity increase. Apart from being different in the sense of techniques for power allocation per subcarrier, various algorithms mutually differ depending on whether optimal power allocation per subcarrier is done for S and RS stations together, or separately — first for S and then for RS, or the power allocation per subcarrier is performed only at the relay station [34–39, 45–48].

Next interesting and very actual solution, for getting full benefits of OFDM and relay techniques combination, deals with the introduction of subcarrier mapping (permutation) at relay stations. At the relay station, subcarriers from the S-RS link are mapped on appropriate subcarriers from the RS-D link, depending on the level of their instantaneous SNRs. This approach has been first considered in [20], with some modifications appeared a couple of months later in [21, 22]. The presented analyses have shown that an OFDM dual-hop relay system, with no direct communication link between the source and the destination, achieves the highest capacity if at RS a subcarrier with the best SNR at the first hop is mapped on a subcarrier with the best SNR at the second hop, a subcarrier with the second best SNR at the first hop on a subcarrier with the second best SNR at the second hop, and so on. This solution has attracted a lot of research interest lately, and a number of results have appeared, investigating performance and possibilities for practical implementation of the proposed scheme.

However, majority of the available research papers and investigations lack a mathematical approach in analyzing performance of OFDM based relay systems with implemented subcarrier permutation (SCP). At the same time, in order to evaluate real effects of additional signal processing introduced through subcarrier permutation at RS and thus to define an optimal OFDM relay system which can be proposed as a potential solution for future WLAN and WWAN networks, it is necessary to develop an analytical model for determination of exact levels of performance parameters that can be achieved in OFDM based relay systems applying different SCP schemes. This is actually the main direction and purpose of the material presented in this book. For its better understanding, the concept of OFDM based relay systems with SCP is described first, with the explanations regarding the possibilities for capacity increase and BER enhancement.

3.3 OFDM Relay Systems with Subcarrier Permutation

As it has already been stated, in obtaining the level of OFDM based relay systems performance, required for high quality wireless data communications, the subcarrier permutation (SCP — *Subcarrier Permutation*) can be incorporated at relay stations. Hence, for the system with two hops, the knowledge about SNRs of all subcarriers at S-RS and RS-D links has to be used at the RS terminal, so that the mapping of subcarriers from the first hop to the subcarriers on the second hop is done, following the chosen metrics. In other words, RS terminals have to be equipped with a special block performing SCP, which uses the estimated information on S-RS channel conditions and the information received from terminal D about the RS-D channel, in the case of half-duplex mode for which reciprocity of RS-D link is valid. Figure 3.6 illustrates a block diagram of such relay station with subcarrier permutation, when AF relaying with fixed gain is applied. In this RS terminal, after getting back into baseband, analog/digital conversion and elimination of a guard interval (GI), OFDM demodulation is done (FFT block). Then, the mapping is performed, i.e., permutation of subcarriers from the first hop to the subcarriers of the second hop, depending on instantaneous SNR for each of the subcarriers. Following SCP, the IFFT block for OFDM modulation is inserted. OFDM signal is then amplified with a fixed gain G, a guard interval is added, and after digital/analog conversion and up conversion, the signal is forwarded towards terminal D.

Information about the performed subcarrier permutation at RS has to be forwarded to terminal D, so that the appropriate signal demodulation will be done. It would be logical to send it in the form of preamble for each OFDM frame. However, for OFDM systems with a high number of subcarriers, this could be problematic due to significant increase of signalization overhead. For example, if OFDM system has 1024 subcarriers, then additional 10 bits for each subcarrier has to be provided within one OFDM frame, or 10240 bits in total have to be sent in order to define the subcarrier position within the OFDM symbol before the permutation. Since, the practical realization of such

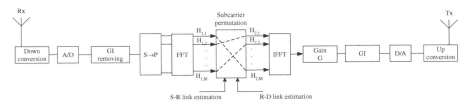

Fig. 3.6   Block diagram of OFDM AF relay station with FG and SCP.

a system would represent a very inefficient solution, the adjacent subcarriers can be grouped in chunks, and then chunk based permutation can be performed, based on the average SNR value on each chunk [22].

It is expected that the above described subcarrier, or chunk, permutation leads towards performance enhancements for the considered OFDM based relay systems. By developing appropriate analytical models, it would be possible to identify exact levels of those performance enhancements, when for example achievable capacity and BER improvements are considered.

3.3.1 Capacity Enhancement in OFDM Relay Systems

The initial idea about the subcarrier permutation for OFDM relay systems has first been presented in [20], and a bit later in [21] and [22]. It has been proved in [22] that dual-hop AF relay system with fixed gain, in the scenario when there is no direct link between terminal S and terminal D, and in the special case with noiseless relay station, the system achieves the maximal capacity if so called Best-to-Best SCP (BTB SCP) is applied. This kind of SCP scheme is realized when the subcarrier with the best SNR from the first hop is mapped to the subcarrier with the best SNR at the second hop, then the subcarrier with the second best SNR from the first hop is mapped to the subcarrier with the second best SNR at the second hop, and so further on. General proof that this kind of SCP scheme maximizes received SNR and achievable capacity in OFDM AF relaying systems is given in [23].

Full comparisons of the average ergodic capacity per subcarrier for dual-hop OFDM AF FG relay system, with and without BTB SCP scheme, are illustrated in Figure 3.7. Graphs for the ergodic capacity are obtained by simulations, assuming Rayleigh narrowband fading statistics on both hops. The case of RS terminal transmitting with the same average power as terminal S is considered, i.e., the gain given with the relation (2.4) is chosen, with $\epsilon_R = \epsilon_S$. Three different scenarios are taken into account: (1) the average SNR at the RS-D link is five times higher than the average SNR at the S-RS link; (2) the average SNRs at the S-RS link and the RS-D link are identical; (3) the average SNR at the RS-D link is five times lower than the average SNR at the S-RS link.

The presented average ergodic capacities per subcarrier values show that, for all SNR values, the BTB SCP scheme enables higher capacity in comparison with the random subcarrier mapping, presented as the system without SCP.

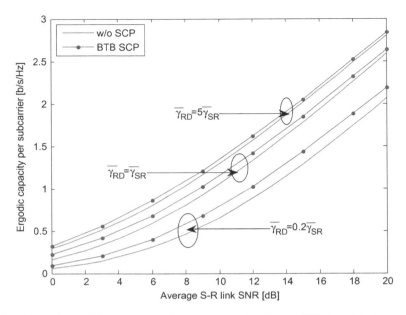

Fig. 3.7 Comparisons of the average ergodic capacity per subcarrier for OFDM AF FG relay systems, with and without BTB SCP.

At the same time, it can be noticed that the mutual difference in the achieved capacities of the systems compared decreases if RS is closer to the destination, i.e., if the average SNR at the RS-D link is higher than the average SNR at the S-RS link. For a given value of the average ergodic capacity over 0,5 b/s/Hz, the BTB SCP scheme enables SNR gain of around 1 dB over the system with no SCP, when the average SNR at the RS-D link is five times lower than the average SNR at the S-RS link. If $\bar{\gamma}_{RD} = \bar{\gamma}_{SR}$, this SNR gain is approximately 0,5 dB, while for $\bar{\gamma}_{RD} = 5\bar{\gamma}_{SR}$, SNR gain becomes only 0,25 dB. For the case $\bar{\gamma}_{SR} = 10$ dB, the system with BTB SCP achieves capacity increase of 18,7% when compared with the relay system with no SCP, for $\bar{\gamma}_{RD} = 0,2\bar{\gamma}_{SR}$. This enhancement comes to 13,3% and 5,4%, when $\bar{\gamma}_{RD} = \bar{\gamma}_{SR}$ and $\bar{\gamma}_{RD} = 5\bar{\gamma}_{SR}$, respectively.

3.3.2   BER Performance Improvement in OFDM Relay Systems

Besides the achievable capacity, the probability of error per bit represents another performance parameter of interest, especially when dealing with applications which demand constant data transmission rates. First analyses focused on subcarrier permutation impact on BER minimization for OFDM AF relay systems can be found in [25], where various communication scenarios in dual-hop AF relay system with three terminals were considered. In the scenario with no direct communication between terminal S and terminal D, by applying majorization theory of inequalities, it has been shown that BER performance of OFDM AF relay system with variable gain can be minimized using the BTB SCP scheme only in the range of small SNR values. Hence, for high SNRs, BER performance can be improved using a modified SCP scheme where the subcarrier with the best SNR from the first hop is mapped to the subcarrier with the worst SNR at the second hop, then the subcarrier with the second best SNR from the first hop to the subcarrier with the second worst SNR from the second hop, and so further on. Such subcarrier permutation scheme is called *Best-to-Worst* **SCP (BTW SCP)**. Thus, it means that the requirement for maximizing the achievable capacity and the requirement for minimizing BER values, might lead to completely different subcarrier permutation schemes at relay stations, which depends on SNRs of subcarriers from both hops.

Figure 3.8 shows simulation results for BER performances of dual-hop OFDM AF VG relay system with BTW SCP and BTB SCP schemes, with no direct communication between the source and the destination. QPSK modulation is assumed and, for the sake of comparison, the results for OFDM AF VG relay system without subcarrier permutation at RS are also illustrated. The presented simulation results are obtained for the case when fading on both hops has Rayleigh narrowband statistics, and the average SNR at the S-RS link is equal to the average SNR at the RS-D link.

The presented BER graphs show that for small values of the average SNR per hop, the best BER results are achieved with the BTB SCP scheme, while for the values of the SNR per hop over 16 dB the BTW SCP scheme achieves the lowest BER values. The difference in BER performances between the two systems is very small, when QPSK modulation and AF VG forwarding are applied. For example, for the BER value of 10^{-1} the relay system with BTB SCP has the SNR gain which is a bit less than 1 dB compared to the system without SCP, and SNR gain of about 1,5 dB compared to the relay system with the BTW SCP scheme. When the average SNR per hop is 25 dB, the relay system with BTW SCP achieves BER value of $3,1 \cdot 10^{-3}$, the relay system

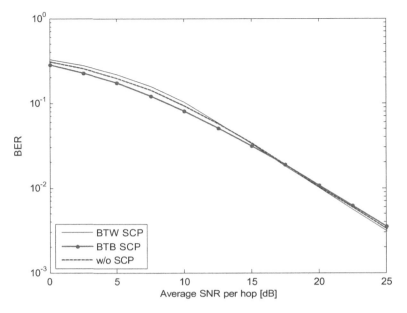

Fig. 3.8 BER performance of QPSK modulated OFDM AF VG relay system, with and without SCP.

without SCP $3,3 \cdot 10^{-3}$, while in the relay system with BTB SCP scheme the BER is equal to $3,5 \cdot 10^{-3}$.

Having in mind the above given explanations, it is clear that in order to achieve BER performance enhancement RS terminal must know SNR boundary value between low and high SNR regions, where it should switch from one SCP scheme to another. This boundary between low and high SNR regions depends on the type of the modulation scheme applied. In the case of dual-hop OFDM AF VG relay system with QPSK modulation, it has been shown [25], that it can be considered for this relay system to operate in the range of high SNR values if for all instantaneous SNRs, for all subcarrier pairs, the following condition is fulfilled:

$$\gamma_{i,SR}^2 \gamma_{j,RD}^2 \geq \psi(\gamma_{i,SR}, \gamma_{j,RD}), \quad 1 < i < M, \ 1 < j < M \qquad (3.4)$$

where:

$$\psi(\gamma_{i,SR}, \gamma_{j,RD}) \triangleq 5(\gamma_{i,SR}^2 \gamma_{j,RD} + \gamma_{i,SR} \gamma_{j,RD}^2) + 2(\gamma_{i,SR}^2 + \gamma_{j,RD}^2)$$
$$+ 4(\gamma_{i,SR} + \gamma_{j,RD}) + 9\gamma_{i,SR}\gamma_{j,RD} + 2 \qquad (3.5)$$

In the above relations, $\gamma_{i,SR}$ denotes SNR on the i-th subcarrier at RS, $\gamma_{j,RD}$ is SNR on the j-th subcarrier at the destination, and M is a number of subcarriers in the OFDM system. On the other side, the considered OFDM AF VG relay system operates in the range of low SNR values, if the instantaneous SNR values for all subcarrier pairs satisfy:

$$\gamma_{i,SR}^2 \gamma_{j,RD}^2 \leq \psi(\gamma_{i,SR}, \gamma_{j,RD}) \tag{3.6}$$

Relation (3.5) which defines the boundary between the low and high SNR regions is obtained by taking partial derivatives of the BER expression for QPSK modulation, over SNRs on the S-R and R-D links:

$$\frac{\partial^2 P_{b,k}(\gamma_{k,end})}{\partial \gamma_{k,SR} \partial \gamma_{k,RD}} = \frac{\partial^2 Q(\sqrt{\gamma_{k,end}/2})}{\partial \gamma_{k,SR} \partial \gamma_{k,RD}} \tag{3.7}$$

$Q(\cdot)$ denotes Q function, which can be defined using complementary error function:

$$Q(x) = 0,5 \, erfc(x/\sqrt{2}) \tag{3.8}$$

where:

$$erfc(x) = \frac{2}{\sqrt{\pi}} \int_x^\infty e^{-t^2} dt \tag{3.9}$$

After some mathematical transformations, it can be shown that the sufficient condition for the considered system to operate in the range of high SNR values is given with:

$$\gamma_{i,SR} \geq \gamma_{high} \quad \text{and} \quad \gamma_{j,RD} \geq \gamma_{high}, \quad \text{for each } i, j \tag{3.10}$$

while, for the range of small SNRs values it comes to:

$$\gamma_{i,SR} \leq \gamma_{low} \quad \text{or } \gamma_{j,RD} \leq \gamma_{low}, \quad \text{for each } i, j \tag{3.11}$$

In the above relations:

$$\gamma_{high} = (5 + \sqrt{33})/2 + \sqrt{17 + 3\sqrt{33}} \approx 11,22 \tag{3.12}$$

$$\gamma_{low} = (5 + \sqrt{33})/2 \approx 5,37 \tag{3.13}$$

The values γ_{high} and γ_{low} are obtained from the equation $\psi(\gamma_{i,SR}, \gamma_{j,RD}) = 0$. It should be stressed out that inequalities (3.10) and (3.11) have to be fulfilled for all subcarrier pairs in order to have OFDM AF VG relay system operating in the region of high, or in the region of low SNR values, respectively. However, this system can be in the region belonging neither to high nor to low SNR regions, if none of the above defined conditions is fulfilled for all subcarrier pairs.

Figure 3.9 shows the boundary between the ranges of low and high SNR values, for the OFDM AF VG dual-hop relay system with QPSK modulation. When instantaneous SNRs for all subcarrier pairs are greater than 11,22 (10,5 dB), then it can be considered that this particular relay system operates in the range of high SNR values. When at least one of the subcarriers in each pair has the instantaneous SNR bellow 5,37 (7,3 dB), then the relay system is located in the range of small SNR values. If those requirements are not fulfilled, the analyzed system is considered to be in so called mixed SNR regime, operating neither in the high SNR region, nor in the low SNR region.

Furthermore, using the relations (3.4) and (3.6), or (3.10) and (3.11), it is not possible to find an exact value of the average SNR per hop where the transition from BTB SCP to BTW SCP scheme has to be done, in order to

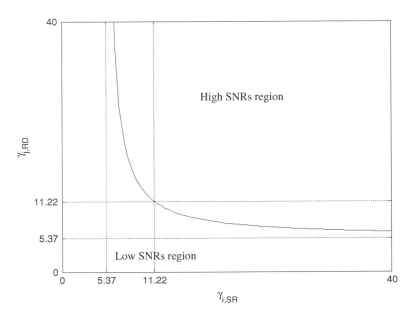

Fig. 3.9 Boundary between ranges of high and low SNRs, for OFDM AF VG relay system (QPSK modulation).

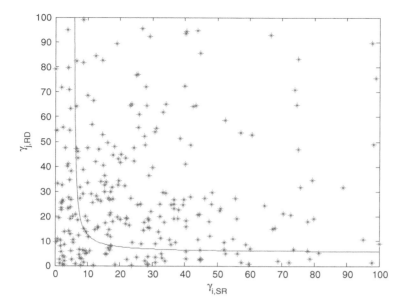

Fig. 3.10 Exponential distribution of instantaneous SNR values $\gamma_{i,SR}$ and $\gamma_{j,RD}$, for $\bar{\gamma}_{SR} = \bar{\gamma}_{RD} = 15\,\text{dB}$.

minimize BER values. This is proved with the Figure 3.10, which gives the instantaneous SNR values for each subcarrier pair of an OFDM AF relay system with 256 subcarriers. Since fading with Rayleigh statistics is assumed for each hop, the instantaneous SNR values follow exponential distribution. For the analyzed relay system with the QPSK modulation and AF VG forwarding, the presented Figure 3.10 gives also a clear illustration of the boundary between the regions of small and high SNR values.

Despite the fact that the average SNR per hop of 15 dB could lead towards expectation that the majority of subcarrier pairs will be in the range of high SNRs, the Figure 3.10 shows that their number is almost equal with the number of subcarrier pairs being in the range of small SNRs. This appears due to exponential distribution, characterizing the instantaneous SNR values of subcarriers exposed to Rayleigh fading, which has high density in the range of SNR values close to zero. That is why for small average SNRs per hop, the majority of subcarrier pairs will be in the range of small SNRs, what is further more proved with the example shown in Figure 3.11, obtained for the case when $\bar{\gamma}_{SR} = \bar{\gamma}_{RD} = 5\,\text{dB}$.

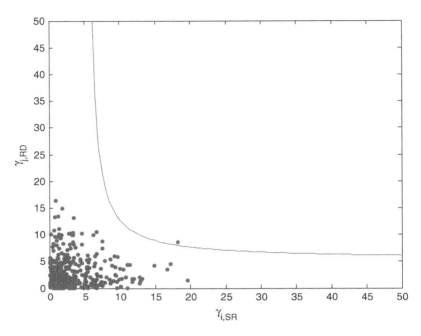

Fig. 3.11 Exponential distribution of instantaneous SNR values $\gamma_{i,SR}$ and $\gamma_{j,RD}$, for $\bar{\gamma}_{SR} = \bar{\gamma}_{RD} = 5\,$dB.

Apart from the above conclusions for the relay system with QPSK modulation and AF VG relay forwarding, the similar analyses can be applied for other relay systems, such as for example BPSK modulated OFDM AF VG and OFDM AF FG. In order to precise their boundaires between the ranges of high and low SNRs, the expression for bit error probability per subcarrier for the case of BPSK modulation is used:

$$P_{b,k} = Q(\sqrt{2\gamma_{k,end}}) \tag{3.14}$$

as well as the relations for the instantaneous SNRs per subcarrier for AF VG system (2.23), so that the partial derivation of BER expression over $\gamma_{k,SR}$ and $\gamma_{k,RD}$ is obtained as:

$$\frac{\partial^2 P_{b,k}(\gamma_{k,end})}{\partial\gamma_{k,SR}\partial\gamma_{k,RD}} = \frac{1}{4\sqrt{\pi}}\frac{e^{-\gamma_{k,end}}}{\sqrt{\gamma_{k,end}}}\frac{2(\gamma_{k,SR}\gamma_{k,RD})^2 - 2(\gamma_{k,SR} + \gamma_{k,RD})}{(1 + \gamma_{k,SR} + \gamma_{k,RD})^4}$$
$$\times \Big[-3\gamma_{k,SR}\gamma_{k,RD} - \gamma_{k,SR}^2 - \gamma_{k,RD}^2 - \gamma_{k,SR}^2\gamma_{k,RD} - \gamma_{k,SR}\gamma_{k,RD}^2 - 1 \Big]$$
$$\tag{3.15}$$

Using the above relation, a function representing the boundary between the ranges of high and low SNRs can be derived in the following form:

$$\psi(\gamma_{i,SR}, \gamma_{j,RD}) \triangleq (\gamma_{k,SR} + \gamma_{k,RD}) + 1,5\gamma_{k,SR}\gamma_{k,RD}$$
$$+0,5(\gamma_{k,SR}^2 + \gamma_{k,RD}^2 + \gamma_{k,SR}^2\gamma_{k,RD} + \gamma_{k,SR}\gamma_{k,RD}^2 + 1)$$

$$(3.16)$$

Having in mind that $\gamma_{k,SR} \geq 0$ and $\gamma_{k,RD} \geq 0$, and by equating the expression (3.16) with zero, the value $\gamma_{high} = 2,414$ (3,82 dB) is obtained, while the condition $\gamma_{k,SR} = \gamma_{k,RD}$ in the (3.16) after is equaled to zero, gives $\gamma_{low} = 1$ (0 dB). Introducing those values into relations (3.10) and (3.11), it becomes possible to define conditions that all subcarrier pairs have to fulfill in order to provide operation of the analysed relay systems in the range of small or high SNR values, respectively.

The obtained boundary between high and small SNR values for the BPSK modulated OFDM AF VG relay system is shown in Figure 3.12. It can be noticed that, in comparison with the QPSK modulated OFDM AF VG relay

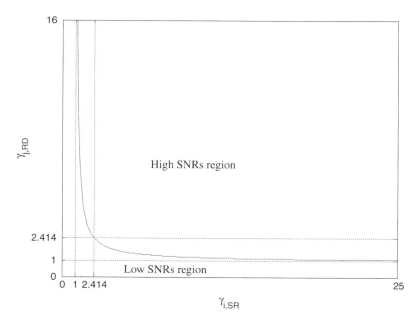

Fig. 3.12 Boundary between ranges of high and small SNRs, for BPSK modulated OFDM AF VG relay system.

system, this boundary is shifted towards lower SNRs. This means that, for the system with BPSK modulation, the BTW SCP scheme represents an optimal solution from the point of BER performance for lower SNR values per hop, when compared with the system with QPSK modulation.

For the BPSK modulated OFDM AF FG relay system, after incorporating the expression for SNR per subcarrier at the receiver (2.8) into BER relation (3.14), its partial derivation gives:

$$\frac{\partial^2 P_{b,k}(\gamma_{k,end})}{\partial \gamma_{k,SR} \partial \gamma_{k,RD}} = \frac{1}{2\sqrt{\pi}} \frac{e^{-\gamma_{k,end}}}{\sqrt{\gamma_{k,end}}} \frac{\rho(\gamma_{k,SR}\gamma_{k,RD} - \gamma_{k,RD} - \rho)}{(\gamma_{k,SR} + \gamma_{k,RD})^3} \qquad (3.17)$$

where ρ is defined with the relation (2.12). In the case of this particular system, the boundary between the ranges of low and high SNR values is defined with instantaneous SNRs for the subcarriers of both hops fulfilling the condition:

$$\gamma_{k,SR}\gamma_{k,RD} = \gamma_{k,RD} - \rho \qquad (3.18)$$

Thus, the value denoted as γ_{high} for BPSK modulated OFDM AF FG relay system is obtained as equal to 2,06 (3,14 dB), while γ_{low} can be defined

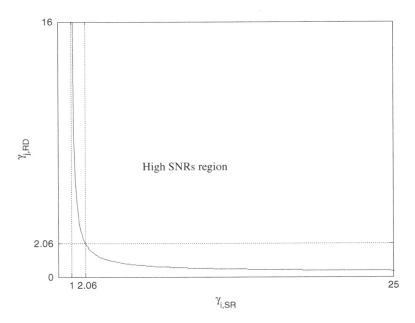

Fig. 3.13 Boundary between ranges of high and low SNRs, for BPSK modulated OFDM AF FG relay system.

only for instantaneous SNRs at the RS-D link, getting the value of 1 (0 dB), (Figure 3.12). Those results show that, in the case of BPSK modulated OFDM AF FG relay system, the transition from the BTB SCP to the BTW SCP scheme, with the purpose of minimizing BER, should be done for lower SNR values, when compared with the BPSK modulated OFDM AF VG relay system.

Following the given explanations regarding the impacts of different subcarrier permutation schemes on the system capacity and BER performance, the considerable benefits related with their incorporation in OFDM based relay systems become obvious. However, in order to evaluate real and precise effects of additional signal processing introduced through subcarrier permutation at RS, it is necessary to develop an analytical model for determination of exact levels of performance parameters that can be achieved in OFDM based relay systems applying different SCP schemes. In that manner, and taking into account the complexity of the overall system as another parameter of interest, it will be possible to define an optimal OFDM relay system which can be proposed as a potential solution for future WLAN and WWAN networks.

4

Relay Stations in Wireless Cellular Networks

4.1 OFDM Relay Systems in WWAN

Operators, which enable services to mobile users over WWAN broadband networks, are faced with constantly raising challenges to provide enough capacity at the network access level, in environments characterized with ever increasing users' demands. This certainly comes with new web based services and multimedia applications, as well as with better affordability of user terminals offering possibilities to access Internet and its applications in different mobility scenarios. One of the common solutions applied in wireless communication systems for increasing their capacity is using the wider channel bandwidth, what often implies a shift toward higher frequency bands. However, this approach causes a number of limitations, like: availability of licensed frequency bands, eventual interference with other systems if unlicensed bands are chosen, reduction in coverage area of base stations due to higher propagation losses at higher frequencies. Another possible solution can be realized through reduction of coverage range of a single base station. This leads to increased number of base stations needed to cover a certain region and, consequently, to doubtful economic profitability, especially in urban areas where a problem already exists in finding a site for the base station and obtaining a backhaul link, which should be realized with optical cable in order to support raising demands for capacity. So far, research works and tests have shown that even the solutions based on advanced technologies (MIMO, signal processing, OFDM as transmission technique at the physical layer) are

OFDM Based Relay Systems for Future Wireless Communications, 59–77.

not sufficient to ensure the required service quality for all users, including those located at the cell edges where the received signal is highly attenuated and interference becomes a serious problem. Therefore, following some of the ideas and results achieved with WLAN networks, OFDM based relaying has been proposed as a logical solution to increase capacity of WWAN networks with the simultaneous extension of theirs coverage areas. The main advantage of this approach with relay stations, when compared with the one implying an increased number of base stations, is related with the cost reduction. Namely, relay stations are connected with a base station over a wireless link, thus eliminating a need for a fixed backhaul connection. Apart from that, since relay stations are less complex than base stations and they transmit with lower power, their price would be accordingly lower and it would be easier to find a place for their mounting. On the other hand, for this solution with OFDM based relaying a certain shortcomings could also be observed, like:

- Increased complexity of the base station which has to incorporate additional *scheduling* functions, in order to support resource allocation for relay stations.
- Greater signaling overhead so that handover procedures, synchronization, security can be controlled.
- Longer end-to-end delay, if DF relaying is applied and even more evident for multi-hop configurations.
- Possibility that the level of interference increases. In the case when relay stations are used for enlarging the coverage area, intercell interference might appear, while in the case when RSs have the role in capacity improvement additional intracell interference could be generated.
- Necessity of a very precise synchronization, especially when relay stations take part in some cooperation scenarios.
- Additional channel estimations when coherent modulations are used.

Despite the above given disadvantages, numerous analyses have shown that benefits achieved using OFDM based relaying techniques are much more important and this technological solution will find its place in future WWAN networks [53, 103, 104].

At the same time, it is worth noticing that WiMAX systems, with the adoption of the IEEE 802.16e standard, have stepped out from the original role of broadband systems for fixed wireless access, and nowadays they offer services to mobile users as well. They are actually in situation to be competitive not only with fixed access technologies providing high quality of services for fixed users, but also with the third generation of mobile cellular systems, which have achieved good coverage with satisfactory service quality, even for high mobility users. Therefore, in order to improve their position and recognize benefits of relay techniques for increasing the system coverage areas and capacity, members of WiMAX Forum have proposed the implementation of OFDM based relay systems. First steps in that direction started in 2006 with the work on the IEEE 802.16j standard, which specifies possibilities for relay stations deployment in WiMAX systems.

Relay stations are also foreseen for the next generation of WiMAX systems, formulated through the IEEE 802.16m standard. Working in the same direction, 3GPP (*Third Generation Partnership Project*) group has adopted relaying techniques in its new generation of mobile cellular systems, known as LTE-Advanced (*Long Term Evolution — Advanced*). Both these standards assume OFDMA (*Orthogonal Frequency Division Multiplexing*) as a physical layer solution for downlink communications. Uplink transmission in IEEE 802.16m will also rely on OFDMA, while in LTE-Advanced systems SC-FDMA (*Single Carrier — Frequency Division Multiple Access*) will be used in order to improve the efficiency of the power amplifier in the user equipment, as it is known that OFDM has a problem with a big peak-to-average power ratio. LTE-Advanced, as well as the IEEE 802.16m systems fulfill IMT-Advanced (*International Mobile Telecommunications — Advanced*) requirements and both are accepted as IMT-Advanced technologies.

4.2 Relay Specifications in IEEE 802.16j Standard

IEEE 802.16j is the first standard for broadband wireless access systems which includes deployment of relay stations [105]. It is adopted in June 2009 under the original title *"Multihop relay specification"*, being actually an addition to the IEEE 802.16-2009 standard in the sense of defining all necessary parameters for including multi-hop communication over relay stations in WiMAX systems. Since the IEEE Working group 16j was established in March 2006, it can be observed that it has accomplished its tasks in relatively short time, having in mind a number of modifications at the physical layer and the MAC (*Medium Access Control*) layer which had to be fully compatible with the IEEE 802.16e standard. Actually, the formulation of the IEEE 802.16j standard avoided new additional functionalities of user terminals which have not already been a part of 802.16e specifications, so that in practice there would be no difference for end-terminals while communicating with a base station or with a relay station. This approach obviously limits possibilities of using all potentials of relay stations, but it has been adopted in order to increase the presence of WiMAX systems based on the 802.16j standard, as many manufacturers of user equipment have already started providing WiMAX compatible mobile terminals. It can be observed that the IEEE 802.16j standard has not been aimed to give specifications for a new cellular network with multi-hop possibilities, but first of all to extend the already existing 802.16 standard in the sense of including multi-hop functionalities.

Possible scenarios for relay stations deployment in 802.16j based networks are defined in [106], and they are illustrated in Figure 4.1.

One of the RS deployment scenarios assumes implementation of fixed relay stations (F-RS), which will be deployed by network operators, taking

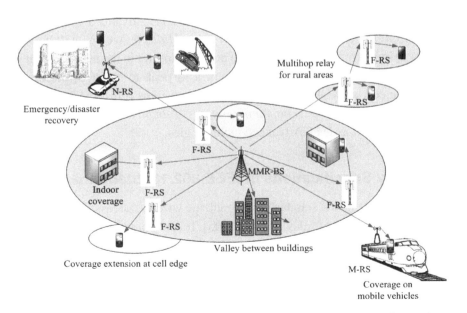

Fig. 4.1 Examples of possible scenarios for relay stations deployment in IEEE 802.16j networks.

into account the following requirements: coverage area enlargement, capacity increase in certain regions with specific demands for quality of service, reduction in total transmitted energy, and/or quality of service improvement at the locations where, so called "coverage holes" appear. These relay stations will be usually installed in a manner to maintain the direct line of sight with the base station (exp. at the buildings' roofs), but they can also operate when there is no direct line of sight with BS.

Another scenario of relay stations incorporation is related with the intention to obtain a better signal coverage inside buildings, since in many practical situations the quality of service for users located in buildings is not satisfactory. Those relay stations (F-RS) can be installed by network operators, locating them as close as possible to the area of interest. They can also be implemented by end users themselves, when a need to cover certain places inside buildings arises. Apart from deploying fixed relay stations, this scenario can include implementation of nomadic relay stations (N-RS) and the operation of relay stations can even be maintained in conditions when there is no direct line of sight with the base station. The mentioned relay stations of nomadic nature are usually used for the temporary coverage of areas where a great number of users is present at the relatively small area for limited period of time (sport events, concerts,...). If more nomadic relay stations are implemented, they can direct the traffic toward base stations which are located in different cells, offering thus better services for the users. Another environment for nomadic RS deployment is the one related with the urgent situations, like natural catastrophes, when they can substitute destroyed fixed elements of telecommunication infrastructure. That is why a special attention is paid to provide their battery power supply.

Figure 4.1 illustrates also a scenario where relay stations are used for better signal coverage in different transportations (trains, buses, ferries, etc). Here, so called mobile relay stations (M-RS) are deployed directly at transportation, whose movement normally causes fast changes from the coverage area of one base station to the coverage area of another base station. Therefore, relay stations are fairly complex and it is expected that a single M-RS serves a number of end users directly or in multi-hop mode through another relay stations, which could be installed for example in different cars of a bigger train.

The above described examples are not exhaustive when scenarios for relay stations deployment in 802.16j networks are concerned. Still, despite the standard full formulation, it is hard to believe that all mentioned relay stations will soon be applied. In initial phases, it is realistic to expect their implementation in cellular systems where traditionally additional base stations have been installed in order to extend coverage range or improve the quality of service. Apart from being less complex, this approach with RSs will be more cost effective [51].

When talking about different scenarios for relay stations deployment in the IEEE 802.16j systems, it is important to mention that those relay stations operate in one of the following two modes: **transparent** or **nontransparent** [105]. In the transparent mode, RS is transparent to a mobile station. A base station transmits directly over downlink synchronization and control channels to a mobile station, while RS forwards the assigned unicast traffic toward the mobile station. In this mode of operation, the maximum number of hops between BS and MS is two and BS and RSs within the same sector work on the same frequency. RSs are not allowed access to security keys or connection identifiers, as all security associations and connections are established between BS and MS. Complete resource allocation in one sector is performed by BS and it is based on the evaluation of effective capacity that can be achieved using a direct BS-MS link and then its comparison with the capacity that can be obtained by forwarding information data to MS over some of RSs present in the sector. If BS decides to forward the data over a particular RS, it sends data to that RS, with the instruction to forward them in the specified part of the frame. Modulation technique and encoding scheme are also chosen by BS, for the BS-RS communication as well as for the RS-MS communication process. When MS changes its location, BS can decide to assign another RS to this mobile station, depending on channels' quality measurements. This is done with no explicit handover procedure, since BS simply sends an instruction to another transparent RS to perform data forwarding in the specified part of the frame. On the other hand, when MS leaves the coverage area of a certain BS, handover is realized in a manner standard for traditional cellular networks. When transparent RSs are applied, it is important to design coverage areas of BSs with the goal to enable satisfactory reception of control information for mobile stations located at the cell edges. Thus, increasing the number of RSs in this mode of their operation, neither the overall coverage area will be extended nor the range of a single BS, but the conditions will be created for the system capacity improvement.

In nontransparent mode of relay stations operation, mobile stations are basically aware of RSs presence in the network. However, in the IEEE 802.16j systems mobile stations can not make a difference between a base station and nontransparent RSs, i.e. mobile stations "see" RS as a base station. Here, BS and RSs in the same sector can operate on the same working frequency, or different frequencies can be used. Also, BS and RS send downlink synchronization and control channels at the same frequency and in the same part of the frame, in accordance with the IEEE 802.16e standard. When a certain MS joins the network, its synchronization can be done with BS or with some of the nontransparent RSs. If it is done with the assistance of RS, then after completing the synchronization process, RS informs BS about the new MS wishing to become a part of the network. After that, RS forwards information, related with security applications and connection, between BS and MS, while keeping connection identifiers for each BS-MS connection established with its participation. Contrary to the situations where RSs operating in transparent mode are deployed, the introduction of nontransparent relay stations enables enlargement of the range and coverage area of a single BS, and/or increase of the system capacity [54].

The IEEE 802.16j standard supports centralized and distributed scheduling modes for resource allocation. In the centralized scheduling mode BS serves as a control node which receives all necessary information about channels and locations of mobile stations. On the base of that, BS performs resource allocation, i.e. decides if information will be directly send to MS or this will be done through a relay station, makes a choice of modulation and encoding processes, etc. This type of centralized scheduling is proposed for applications in cells equipped with transparent relay stations. On the other side, distributed scheduling mode assumes that RSs can decide to which mobile station information should be forwarded, knowing the state of the channels toward users and using medium access control algorithm based on the competition concept. When a network cell incorporates nontransparent relay stations, both scheduling modes could be used. Generally speaking, the main difference between them is in the fact that the centralized mode requires greater signalization overhead, thus enabling better system performance [49].

It is worth mentioning that the IEEE 802.16j standard includes all operations present in 16e protocols, such as: hybrid automatic repeat request (HARQ), request for bandwidth, connection management, sleep and idle mode. Apart from defining basic operations for relay stations working in transparent and nontransparent mode and the structure of frame, the IEEE 802.16j standard gives also a specification for advanced relay techniques. Thus, cooperative communications* between a base station and relay stations is supported, with the possibility to create virtual MIMO system for transmitting information toward mobile station. The standard has additional specifications related with handover procedures for RSs and associated mobile stations. A concept of establishing a group of RSs is also enabled, so that more RSs can be combined to form a group which behaves as a single relay station with respect to the mobile station [54]. Despite the fact that the IEEE 802.16j standard supports the mentioned advanced procedures, it is not expected that in the first phases of relay station implementation in WiMAX systems all of them will be fully exploited.

4.3 Relay Solutions in IMT-Advanced Relay Systems

Standardization Sector of the International Telecommunication Union (ITU-T) announced in March 2008 a call for proposals for radio access technologies of IMT-Advanced systems, i.e. for wireless broadband systems of the next generation, usually referred as 4G. Proposed technologies were supposed to fulfill specified requirements, as for example to enable maximum data rates on downlink of 1 Gb/s for the users with low mobility (up to 10 km/h) and maximum data rates of 100 Mb/s for the users with high mobility (up to 350 km/h). Some of the other technical requirements specified for IMT-Advanced systems are given in Table 4.1 [107].

The call was closed in October 2009, with two technologies being proposed as candidates for IMT-Advanced systems:

- LTE-Advanced, proposed by the 3GPP group, and
- IEEE 802.16m, proposed by the IEEE.

Table 4.1. Requirements for IMT-Advanced systems.

	Downlink	Uplink
Bandwidth	5–20 MHz; optional 40 MHz	
Peak spectral efficiency	15 b/s/Hz	6,75 b/s/Hz
System efficiency*	2,6 b/s/Hz/cell	1,8 b/s/Hz/cell
VoIP capacity	40 users/MHz/cell	
Delay	10 ms (U-plane)/100 ms (C-plane)	

*micro-cell (urban environment)

Both proposed candidates have fulfilled all IMT-Advanced requirements, and have even outperformed many of them. For example, the LTE-Advanced system has reached spectral efficiency of 30 b/s/Hz on downlink, when 8×8 MIMO (*Multiple Input Multiple Output*) configuration is used, and spectral efficiency of 15 b/s/Hz on uplink for 4×4 MIMO configuration [108]. The IEEE 802.16m system achieves spectral efficiency of 16,96 b/s/Hz (TDD mode) on downlink already for 4×4 MIMO configuration, and of 9,22 b/s/Hz (TDD mode) on uplink for 2×4 MIMO configuration [109]. In meeting the requirements set for the IMT-Advanced systems, the proposed technologies incorporate many advanced techniques, with the following ones being common for both of them:

— OFDMA as transmission and multiple access technique at downlink, including the possibility of carrier aggregation, so that up to five 20 MHz channels could be assigned to a single user. Those 20 MHz wide channels are not necessarily adjacent channels, but they can be randomly allocated in the available spectrum, all in order to achieve as better as possible performances.

— Advanced MIMO transmission techniques. A spatial multiplex of up to 4 independent data streams, or a transmit diversity, can be realized at uplink, while on downlink this can be achieved for up to 8 data streams. The most important benefits from the MIMO system incorporation are expected through implementation of multiuser MIMO (MU-MIMO) techniques, which represent a combination of *beamforming* and spatial multiplexing.

— Coordinated multi-point transmission (CoMP), intended for overcoming problems with the quality of service for the users located near the cells edges. CoMP assumes that adjacent base stations coordinate their transmissions toward the users placed near the joint cells' edge, with the goal to reduce the interference. CoMP can also be used for the simultaneous data transmission from the two base stations to a user, or for the combining of the signals received by two base stations from one user.

— Relay systems as a solution for extending the range of communication and for increasing capacity, i.e. for improving the quality of signal coverage within a single cell.

The relay solutions in both accepted IMT-Advanced systems have reduced many of the relay functionalities defined in the IEEE 802.16j standard. The reason was relatively short time period between the announced Call for IMT-Advanced candidate proposals and the deadline for proposals' submission, so the goal was to eliminate all ambiguities that have limited wider implementation of the IEEE 801.16j based systems and to achieve a completely clear system proposal whose implementation in practice will be straightforward. The most important difference between the earlier accepted first OFDM based relay standard, IEEE 802.16j, and the accepted solutions for the IMT-Advanced systems, is that the multi-hop capability is not included in the IMT-Advanced systems, but only two-hop relaying. This significantly simplifies relaying functionality, but it does not mean that some of the future specifications for LTE-Advanced systems or WirelessMAN-Advanced systems will not include this capability, as current research efforts are already conducted in this direction. Besides, it is very important that neither of the accepted IMT-Advanced systems comprises transparent relay stations from the IEEE 802.16j standard, which are denoted in 3GPP as Type II relay stations. For the reminder, transparent relay stations do not have the authority to manage the resources, but they share cell IDs and cell control messages with their serving BSs. Like with the feature of the multi-hop capability, it can be also expected that the transparent (Type II) relay stations will be a part of the upcoming amendments (Releases) of the accepted IMT-Advanced systems due to their possibility to further increase the system capacity, which is proven through numerous theoretical and practical research on their performances in different scenarios. Another "downgrade" feature of the both IMT-Advanced solutions compared to the IEEE 802.16j standard, deals with the exclusion of mobile relay stations (M-RS). Namely, it is foreseen that only the infrastructure based relay stations, placed at the locations chosen be network operators, will be implemented in the first phase of the IMT-Advanced system implementation.

From the description of the relaying concepts in the IMT-Advanced systems given above, it is obvious that they represent only a subset of relay solutions existing in the IEEE 802.16j standard. Figure 4.2 gives an overview of the relations among various relay categories included in three standardized OFDM based relay systems.

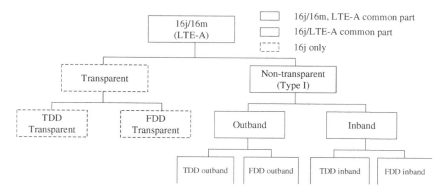

Fig. 4.2 Relay categories in IEEE 802.16j/m and LTE-A systems.

Regarding the abbreviations and nomenclature in Figure 4.2, TDD denotes Time Division Duplex and FDD is Frequency Division Duplex. Outband relaying assumes that RS uses different carrier frequency for transmitting the signal it receives from BS or from MS, while inband relaying assumes that RS transmits the received signal on the same frequency.

4.3.1 Relay Specifications in LTE-Advanced Systems

The LTE Release 10, also known as the LTE-Advanced standard, does not specify a new radio access technology, but it gives the extension of the LTE with, among other new features, support for relaying functionality. Being an evolution of the LTE, Release 10 includes all the features of Release 8 and 9, which will enable operators to perform smooth introduction of the new technology, without significant capital investments. Hence, operators can deploy a LTE network and later can upgrade to LTE-Advanced network at the specific sites where it is necessary. Most of the LTE-Advanced features can be introduced in the network in the form of simple software upgrades. However, relay station deployment assumes one of the biggest changes in the LTE evolution toward LTE-Advanced networks.

As it is earlier described, RS in the LTE-Advanced system will form its own cell, separated from the BS donor cell, and from a mobile station (MS) point of view there will be no difference in communication with RS or with BS. The cell will have its own physical cell ID, and RS will transmit its own synchronization channels, reference symbols, and so on. MS shall receive scheduling information and HARQ (*Hybrid Automatic Repeat Request*) feedback directly from RS and it sends its control channel to RS. Actually, the relay station is low power BS wirelessly connected to the remaining part of the network. In this way, the backward compatibility is provided for mobile stations intended for the LTE systems, specified through Release 8 and Release 9.

The LTE Release 10 specifications have foreseen possibility of inband or outband relaying. Outband relaying will enable simultaneous BS-to-RS and RS-to-MS communication, while inband relaying will rely on half-duplex RS transmission, due to interference caused by RS transmitter to its own receiver. Separation between transmitted and received signals at RS may also be provided through well isolated antenna structures, but this concept should be further explored in practice. Similarly, RS may not be enabled to receive transmission from MS simultaneously while it transmits to BS. In the LTE-Advanced systems a basic scheduling unit is a subframe, which is 1ms long, and 10 subframes constitute a radio-frame. A gap in RS-to-MS transmissions, to allow the reception of BS-to-RS transmissions, is created using MBSFN (*Multicast-broadcast single-frequency network*) subframes. These subframes are present in Release 8 specifications, and they were originally intended for broadcast support, but later have been seen as a generic tool. In an MBSFN subframe the first one or two OFDM symbols in a subframe are transmitted as usual, carrying cell-specific reference signals and downlink control signaling, while the rest of an MBSFN subframe is not used and can therefore be used for the BS-RS communication. Similar to the downlink gaps obtained through the use of MBSFN subframes, there is a need to create gaps in the MS-to-RS transmission in order to achieve the transmission from RS to BS. This is handled by not scheduling MS-to-RS transmissions in some subframes.

The LTE Release 10 defines a new control channel (R-PDCCH — *Relay Physical Downlink Control Channel*) to provide control signaling from BS to RS, since RS can not receive normal control signaling from BS while transmitting cell-specific reference signals in the first part of a MBSFN subframe [112]. This control channel carries downlink scheduling assignments and uplink scheduling grants in the same way as the normal control signaling. Two schemes for R-PDCCH and traffic channel multiplexing are supported: time-division multiplexing + frequency-division multiplexing (TDM+FDM),

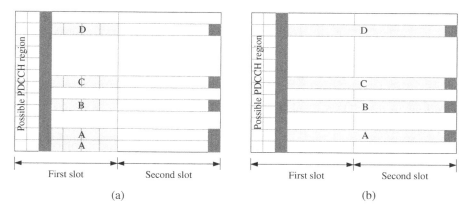

Fig. 4.3 Downlink subframe structure of Type I relay in LTE-A systems, (a) TDM+FDM, (b) FDM.

and pure FDM (Figure 4.3). The start of the R-PDCCH is fixed to the fourth OFDM symbol in the first time slot.

Since R-PDCCH is positioned in the first slot of the subframe, the hybrid TDM+FDM scheme provides shorter decoding latency than the pure FDM scheme. Besides, TDM+FDM offers more frequency diversity when resources are distributed, and more frequency selective or precoding gain when resources are localized. On the other side, pure FDM provides more scheduling flexibility.

Half-duplex inband relaying puts several restrictions and challenges particularly on the physical layer transmission and relay behavior:

— Since RS is basically a forwarding node, the additional delay caused by the relay operation should be as small as possible.
— RS is acting as a multiplexer for uplink traffic and a demultiplexer for downlink traffic. As a consequence, RS-to-BS link can be a bottleneck, and should operate with high spectral efficiency.
— A half-duplex relay continuously switches its radio frequency circuitry between backhaul and access link operations. Switching should be as fast as possible, since a longer switching period automatically means a loss of physical resources.

A major challenge in the backward compatibility requirement for RS arises from the MS's assumption that the downlink cell-specific reference signals

(CRS) are present in each subframe. CRS are used not only for demodulation and channel state information feedback, but also for mobility measurements and radio link monitoring. Therefore, RS has to transmit CRS in each downlink subframe, while still having opportunities to communicate with BS.

The description given above provides basic features overview of relay solutions in the LTE-Advanced systems. There are many open issues which are, and which will be subject of the research activity in the following period, all in order to prepare new relaying solutions that have not find its place in the Release 10. Some of these issues include:

— cooperative transmission,
— multi-hop relaying,
— mobile relay stations,
— power saving options,
— solutions for performance enhancements,
— transparent relay stations (Type II) for capacity enhancement (Figure 4.4), etc.

OFDM based transparent relaying has been well examined through different approaches, in different possible scenarios. Thus, it was expected that the transparent RSs (Type II) will be a part of the 3GPP Release 10, as a

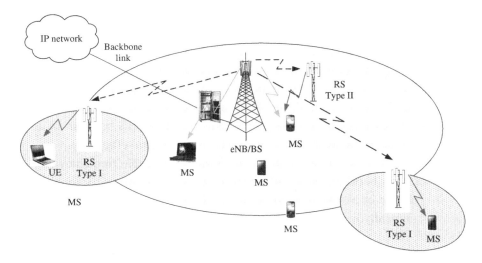

Fig. 4.4   Type I and Type II relay stations in future LTE-Advanced systems.

solution for capacity enhancement. As originally intended, Type II relay station should improve quality of service (i.e., overall capacity) for the mobile station which is located within the coverage area of the base station (eNB), in the scenario including a direct communication between them. This type of relay station should not send any control signals, and it should be actually used for achieving multipath diversity [49]. Both AF and DF based signal processing methods at RS were analyzed as possible solutions for the Type II relay stations, and pros and cons of these methods still represent a topic of interest. In accordance with the predicted functions, the centralized scheduling mode should be necessary for the relay station Type II, while for the Type I relay station both distributed and centralized scheduling modes for resource allocation could be applied. As it is already mentioned, the proven benefits of the Type II RSs implementation represent a strong anchorage for the expectation of their incorporation in future 3GPP Releases.

4.3.2 Relay Specifications in IEEE 802.16m Standard

Relaying in IEEE 802.16m is performed using a decode-and-forward signal processing at RS, which is denoted in this system as advanced relay station (ARS). ARS is fixed, two-hop and non-transparent relay with distributed scheduling. Both time-division duplex (TDD) and frequency-division duplex (FDD) modes are supported. When implementation of relay techniques is concerned, in contrary to the solutions described previously for the IEEE 802.16j standard, the proposed 802.16m standard assumes that mobile terminals are constantly aware of relay stations' presence in the network. In that manner, the relay stations are in a position to perform their role much more efficiently, enabling considerable reduction of interference among the signals sent by BS and relay station [54].

ARS controls the cells on its own with a wireless backhaul connection to the advanced service network gateway (ASN-GW) through the advanced base station (ABS). ARS may control one or several sectors, and a unique physical layer cell identity is provided in each sector controlled by the relay. ABS notifies ARSs and advanced mobile stations (AMSs) of the frame structure configuration. The radio frame is divided in the access and relay zones. In the access zone ABS and ARS transmit to, or receive from, AMS. In the relay zone ABS transmits to ARSs and AMS, or receives from ARSs and AMSs.

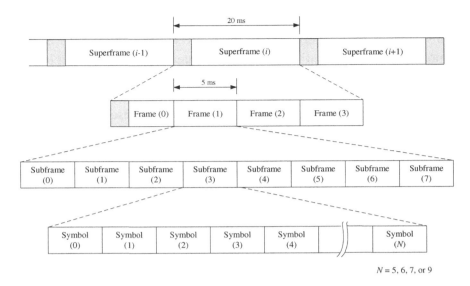

Fig. 4.5   Frame structure in the IEEE 801.16m system.

The 802.16m radio frame structure is presented in Figure 4.5, and it is applied to both FDD and TDD schemes.

The superframe lasting 20 ms is divided into four equally-sized frames, each consisting of eight subframes, which are assigned for either downlink or uplink transmission. The access zone of the radio frame always precedes the relay zone position. Due to asymmetric downlink and uplink traffic load, durations of the access and relay zones could be different in downlink and uplink directions. ABS transmits additional MIMO midambles in downlink relay zone for an operational ARS to perform synchronization with ABS.

The backward compatibility issue in IEEE 802.16 systems is not as successfully solved as it is in the 3GPP group specifications. This is due to intricate protocol design for the IEEE 802.16j standard, which was the reason that simplicity and explicitness in designing the IEEE 802.16m relay have become the general consensus for the purpose of speeding up the implementation. Figure 4.6 shows the interoperability in a complete 802.16m system that consists of an 802.16e MS, an 802.16j mobile relay (MR) — BS/RS, and 802.16m ABS/ARS/AMS. Only the interfaces involving relays are shown in Figure 4.6. It can be seen that the 802.16m does not support backward compatibility with the 802.16j system, except for the 802.16e MS.

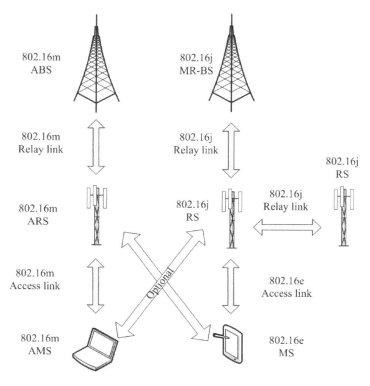

Fig. 4.6 Compatibility among different IEEE 802.16 standards.

It could be assumed that 802.16m systems and 802.16j systems are not compatible, as many options existing in the 802.16j are not included in the 802.16m standard. For instance, mobile relays are not considered as an option for 802.16m systems, as the designs of orthogonal preamble assignment within a cell, hierarchical handover procedures, and so on, undoubtedly make mobile relay much more complicated than fixed relay. Moreover, imposing restriction to two-hop relaying topology eliminates controversial issues of path management and data forwarding. Considering the round-trip delay of conveying control signals by multi-hop relays, the distributed control/scheduling mode adopted in the 802.16m standard seems more efficient than the centralized control/scheduling mode from 802.16j systems.

Like in LTE-Advanced systems, there are still many open issues and a lot of space for performance improving when OFDM based relay systems in IEEE 802.16m standard are considered. Thus, it can be expected that the research topics in the following period will be focused on mobility, power saving, multihop architecture, transparent relaying, cooperative transmission, etc.

4.3.3 Comparisons of IMT-Advanced Relay Systems

Relay solutions for the LTE-Advanced and WirelessWAN-Advanced (802.16m) systems have many similarities, what can be seen from their descriptions given in the former text, but they also differ in a few very important aspects.

The two adopted IMT-Advanced standards assume OFDM based dual-hop relaying, with fixed relay stations implementing signal processing based on the decode-and-forward paradigm. Both of these relay systems apply distributed radio resource scheduling and hop-by-hop HARQ mechanism. Also, the security issues are maintained hop-by-hop, i.e., a relay station (ARS) directly maintains security context with MS (AMS), and at the same time maintains separate security context with BS (ABS).

On the other side, one of the main differences between the two accepted IMT-Advanced systems is in the fact that the LTE-Advanced systems support both inband and outband relaying, while relay stations in 802.16m will implement only inband relaying. That is, the LTE-Advanced Release 10 relay provides more deployment options for operators. Additionally, the difference is in the fact that 802.16m AMS is ARS-aware, whereas the LTE Release 8 user equipment was not allowed to be enhanced to be RS aware. Hence, communications on the BS-to-RS link of the LTE-Advanced inband relay are less flexible and efficient than the appropriate link in 802.16m systems. The difference between these two systems exists also in the frame structure. Namely, AMS is fully aware of the ARS operation within the same cell. Therefore, unlike in the LTE-Advanced system with Type I relay whose BS-to-MS downlink communications are restricted in the MBFSN subframe, regular downlink and uplink subframes are configured by ABS to communicate with ARS. Few enhancements are required for these subframes, mainly at the physical layer to facilitate ARS operations.

Although the standards for IMT-Advanced systems (LTE-Advanced and WirelessMAN-Advanced) have been accepted, an intensive research activity is still ongoing on OFDM based relay systems, with the objective to further improve their performances or to optimize their functions. As it has been already mentioned, one of the interesting solutions for performance improvement is implementation of subcarrier permutation (SCP) at the RS station. This solution may enhance the capacity and/or improve the BER performances. The complete analytical and simulation performance evaluation of different OFDM relay systems with SCP has not been published yet, thus, the general insight in justification of implementation of such a solution has not be provided. As our research in this direction has led to results that are in some areas the first published analytical performance analysis of OFDM based relay system with SCP, we wanted to give comprehensive overview of the benefits of SCP implementation in OFDM relay systems, as well as to define the optimal solution for the future wireless communication systems. In the following Chapters are the results we obtained in our research.

5

Performance of OFDM AF FG Relay Systems with Subcarrier Permutation

It has been already shown in Chapter 3 that subcarrier mapping implemented at relay stations, can improve performances of OFDM based relay systems. Thus, depending on the levels of signal-to-noise ratios at S-RS and RS-D links, better BER and/or capacity values can be obtained. Identification of precise amounts for those performance improvements is necessary, in order to justify SCP incorporation at relay stations. At the same time, it will enable determination of a particular type of OFDM based relay system with SCP which achieves the best performance in given conditions.

This and the next two chapters deal exactly with those essential issues from the point of possible implementation of OFDM relay systems with SCP in the next generation WWANs. They give the complete analytical model for BER and capacity determination for OFDM AF and OFDM DF relay systems with SCP. The closed form relations for the considered performances are derived, enabling comprehensive comparisons of different analyzed systems. On that basis, an attempt to identify optimal solutions for a given set of conditions is also presented.

The description in the above given terms starts with OFDM AF FG relay systems with SCP, what is the subject of this particular chapter. In order to obtain analytical expressions for the BER and achievable capacity for these systems, the statistics relevant for end-to-end performance of wireless systems with relays is first introduced. Thus, adequate relations for the probability

OFDM Based Relay Systems for Future Wireless Communications, 79–105.
© 2012 *River Publishers. All rights reserved.*

density function (PDF) of the system end-to-end SNR with BTW SCP or BTB SCP, are derived. Using the ordered statistics of random variables, those PDFs are presented and then subsequently the received SNR moment generating functions (MGF — Moment Generating Function) are found in [28] and [29]. Using the MGF approach, it becomes possible to derive BER expressions for various modulation schemes (DPSK, BPSK, m-QAM) applied. Similarly, analytical relations for the achievable capacity for each of the considered systems can be obtained. Generally, all analytical results presented are verified by appropriate simulations, proving the sustainability of the idea to use them for comprehensive analyses of OFDM AF FG relay systems with SCP [29].

5.1 System Description

In the analyses presented here, an OFDM dual-hop relay system is considered, comprising: a source terminal S, a half-duplex relay station (RS) and a destination terminal D. It is assumed that all terminals are equipped with single antenna and that there is no direct link between S and D, so that all communications are realized over RS. As it has been already described, in OFDM AF FG relay systems with subcarrier permutation the relay station performs OFDM demodulation (FFT — *Fast Fourier Transformation*) of the signal received from the source, then it applies the permutation of subcarriers in accordance with their channel transfer functions, and then the obtained signal is again OFDM modulated (IFFT — *Inverse Fast Fourier Transformation*). Before its retransmission toward D, the signal is amplified in a block introducing a constant gain. Figure 5.1 illustrates a simplified block scheme of the OFDM AF FG relay station (terminal), which incorporates SCP.

In the analyzed scenario, an ideal estimation of S-RS and RS-D links is assumed, i.e., it is assumed that RS fully knows transfer functions of subcarriers on both hops and that D knows the permutation function applied at the relay station. All terminals are considered to be ideally time and frequency synchronized.

Signal at terminal RS, on the i-th subcarrier after the FFT block, can be represented in the following form:

$$Y_{R,i} = X_{1,i} H_{1,i} + N_{1,i}, \quad 1 \leq i \leq M, \tag{5.1}$$

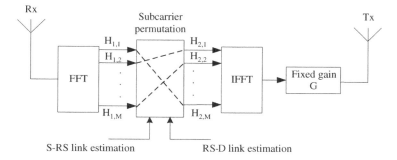

Fig. 5.1 Block scheme of AF RS with fixed gain and SCP.

where M is the total number of subcarriers, $H_{1,i}$ is the channel transfer function of the i-th subcarrier on the S-RS link, and X_i is data symbol transmitted by S on the i-th subcarrier. $N_{1,i}$ denotes an additive white Gaussian noise on the i-th subcarrier at terminal RS, whose variance is $\mathbf{E}(|N_{1,i}|^2) = \mathcal{N}_{01}$ with $\mathbf{E}(\cdot)$ being expectation operator. Assuming that, at relay station, the SCP function $\upsilon(i)$ performs mapping of the i-th subcarrier of the S-RS link to the k-th subcarrier of the RS-D link, then the signal at the destination is given, in the frequency domain, with:

$$
\begin{aligned}
Y_{D,k} &= G H_{2,k} Y_{R,\upsilon(i)} + N_{2,k} \\
&= G H_{2,k} H_{1,i} X_i + G H_{2,k} N_{1,i} + N_{2,k}, \quad 1 \leq k \leq M \quad (5.2)
\end{aligned}
$$

with $H_{2,k}$ being the channel transfer function of the k-th subcarrier on the RS-D link, and $N_{2,k}$ is an additive white Gaussian noise on the k-th subcarrier at the destination, whose variance is $\mathbf{E}(|N_{2,k}|^2) = \mathcal{N}_{02}$. Using the relation (5.2), the instantaneous SNR of the k-th subcarrier at terminal D, can be defined with as:

$$
\gamma_{k,end} = \frac{G^2 \mathbf{E}\{|X_{1,i}|^2\} |H_{1,i}|^2 |H_{2,k}|^2}{\mathcal{N}_{02} + G^2 |H_{2,k}|^2 \mathcal{N}_{01}} = \frac{\frac{\mathbf{E}\{|X_{1,i}|^2\} |H_{1,i}|^2}{\mathcal{N}_{01}} \frac{|H_{2,k}|^2}{\mathcal{N}_{02}}}{\frac{|H_{2,k}|^2}{\mathcal{N}_{02}} + \frac{1}{G^2 \mathcal{N}_{01}}}. \quad (5.3)
$$

If ϵ_S denotes the energy of the symbol transmitted by S and ϵ_R is the energy of the symbol transmitted by RS, then $\gamma_{i,SR} = \epsilon_S / \mathcal{N}_{01}$ represents signal-to-noise ratio of the i-th subcarrier on the S-RS link, and $\gamma_{k,RD} = \epsilon_R / \mathcal{N}_{02}$ is the signal-to-noise ratio of the k-th subcarrier on the RS-D link. In accordance with that, the relation (5.3) becomes:

$$\gamma_{k,end} = \frac{\gamma_{i,SR}\gamma_{k,RD}}{\gamma_{k,RD} + \rho}, \qquad (5.4)$$

where ρ is the coefficient related with the gain applied at the relay station, defined with the relation (2.2.12).

The analyzed scenario assumes no direct communication link between the source and destination terminals, what could be considered, from the point of achievable performance levels, to be a worse case in comparison with the situation when such direct communication exists. The chosen scenario implies subcarriers with independent and identical (i.i.d. — *independent identically distributed*) Rayleigh fading distributions on both hops, so that the probability density function (PDF) of SNR for any of the M subcarriers can be represented as [102]:

$$f_{SR}(x) = \lambda_{SR}\exp(-\lambda_{SR}x), \qquad (5.5)$$

with λ_{SR} denoting an inverse value of the average SNR on the S-RS link, i.e.:

$$\lambda_{SR} = \frac{1}{\overline{\gamma}_{SR}}. \qquad (5.6)$$

The cumulative distribution function (CDF — Cumulative Distribution Function) of SNR for any of the M subcarriers on the S-RS link, becomes:

$$F_{SR}(x) = 1 - \exp(-\lambda_{SR}x). \qquad (5.7)$$

Thus, the first step in presenting the full mathematical model for performance determination of the analyzed OFDM AF FG relay system with SCP, is related with derivation of adequate functions (PDF, CDF, MGF) describing statistics of end-to-end signal-to-noise ratio at its receiving end.

5.2   Statistics of the End-To-End SNR

The relation (5.4), which defines the received signal-to-noise ratio of the k-th subcarrier for OFDM AF relay system with fixed gain and SCP, shows that its statistics (PDF, CDF, MGF) can be determined applying the ordered statistics of random variables. Namely, analyses of all considered OFDM relay systems with SCP are significantly facilitated if the following model is assumed: first terminal RS sorts subcarriers from the S-RS link according to their instantaneous SNR values, then the same terminal sorts in the same manner subcarriers from the RS-D link, and finally performs mapping of the sorted subcarriers from the incoming link to the sorted subcarriers from the outgoing link. Thus, if the BTW SCP scheme is applied, subcarriers from the S–RS link are sorted in increasing order in respect to theirs instantaneous SNRs and then they are mapped to subcarriers from the RS-D link, which are sorted in decreasing order in respect to theirs instantaneous SNRs. For the BTB SCP scheme, subcarriers on both links are sorted in increasing order in respect to theirs instantaneous SNRs.

5.2.1 Ordered Statistics of Random Variables

For the analyzed dual-hop OFDM AF FG relay system with SCP, $f_{k,SR}^{w}(\cdot)$ denotes the PDF of the received SNR for the k-th "worst" subcarrier on the S-RS link, in the sense that, out of the total M subcarriers, it has the k-th lowest SNR. Using an example given for the ordered statistics of random variables given in [102], $f_{k,SR}^{w}(\cdot)$ can be represented with:

$$f_{k,SR}^{w}(x) = M \binom{M-1}{k-1} f_{SR}(x)(F_{SR}(x))^{k-1}(1 - F_{SR}(x))^{M-k} \qquad (5.8)$$

where (:) denotes binomial coefficients. Introducing relations (5.5) and (5.7) into relation (5.8), it is obtained:

$$f_{k,SR}^{w}(x) = M \binom{M-1}{k-1} \lambda_{SR}(1 - e^{-\lambda_{SR}x})^{k-1} e^{-\lambda_{SR}x(M-k+1)}. \qquad (5.9)$$

Applying the binomial expansion:

$$(1 - e^{-\lambda_{SR}x})^{k-1} = \sum_{i=0}^{k-1}(-1)^i \binom{k-1}{i} e^{-\lambda_{SR}xi}, \qquad (5.10)$$

the equation (5.9) is modified as:

$$f_{k,SR}^{w}(x) = \sum_{i=0}^{k-1} \lambda_{SR}\alpha_i e^{-\beta_i \lambda_{SR}x}, \qquad (5.11)$$

while the coefficients α_i and β_i are given with:

$$\alpha_i = (-1)^i M \binom{M-1}{k-1} \binom{k-1}{i}, \tag{5.12}$$

$$\beta_i = i + M - k + 1. \tag{5.13}$$

In order to define the PDF of the end-to-end SNR of OFDM AF FG relay systems with the BTW SCP scheme, it is necessary to know the statistics of decreasingly ordered random variables, since it is relevant for the situation when subcarriers are ordered from the one having the highest SNR to the one having the lowest SNR. Using the similar approach as in the case of increasingly ordered random variables [102], the PDF of decreasingly ordered random variables can be derived. The result is obtained in the form of the PDF of the k-th "best" subcarrier on the RS-D link $f^s_{k,RD}(\cdot)$, in the sense that it has the k-th highest SNR out of the total M subcarriers:

$$f^s_{k,RD}(x) = M \binom{M-1}{k-1} f_{RD}(x)(F_{RD}(x))^{M-k}(1 - F_{RD}(x))^{k-1}. \tag{5.14}$$

$f_{RD}(x)$ and $F_{RD}(x)$ denote the PDF and CDF of the received SNR on the RS-D link, respectively, given with the relations (5.5) and (5.7) where the average SNR on the RS-D link is introduced. For the assumed scenario with subcarriers on the RS-D link with i.i.d. Rayleigh fading, $f^s_{k,RD}(\cdot)$ can be obtained in the form:

$$f^s_{k,RD}(x) = \sum_{i=0}^{M-k} \lambda_{RD}\delta_i e^{-\varepsilon_i \lambda_{RD} x}, \tag{5.15}$$

while coefficients δ_i and ε_i have the following values:

$$\delta_i = (-1)^i M \binom{M-1}{k-1} \binom{M-k}{i}, \tag{5.16}$$

$$\varepsilon_i = i + k. \tag{5.17}$$

Thus, knowing the PDF of SNR for the k-th best subcarrier on the S-RS link, (5.11), as well as the PDFs of SNR for the k-th best or k-th worst subcarriers on the RS-D link, (5.15), it becomes possible to derive closed form expressions for the PDF of the end-to-end SNR for OFDM AF FG relay systems with BTB SCP or BTW SCP.

5.2.2 PDF of SNR for BTW SCP Scheme

Assuming that, according to theirs instantaneous SNRs, subcarriers on the S-RS link are increasingly ordered at terminal RS, and then mapped to decreasingly ordered subcarriers of the RS-D link, the PDF of the end-to-end SNR for relay systems with BTW SCP scheme can be derived in a following way. First of all, instantaneous SNR at the output of the FFT block in terminal D, can be defined with:

$$\gamma_{k,end} = \frac{\gamma_{k,SR}\gamma_{k,RD}}{\gamma_{k,RD} + \rho} = \frac{\gamma_{k,SR}}{1 + \rho z_k}, \tag{5.18}$$

where $z_k = 1/\gamma_{k,RD}$. Knowing that the PDF of the random variable $\gamma_{k,RD}$, for the assumed scenario of decreasingly ordered Rayleigh fading subcarriers on the RS-D link, has the form given with the relation (5.15), the PDF of the new random variable z_k can easily be found. For that purpose, the CDF of the random variable z_k is first defined, using the already known CDF of the random variable $\gamma_{k,RD}$. Denoting with $P(\cdot)$ the event probability, it can be written:

$$F_{z_k}(Z) = P(z_k \leq Z)$$
$$= P\left(\frac{1}{\gamma_{k,RD}} \leq Z\right) = P\left(\gamma_{k,RD} \geq \frac{1}{Z}\right) = 1 - F_{\gamma_{k,RD}}\left(\frac{1}{Z}\right). \tag{5.19}$$

From the above expression, the PDF of the random variable z_k is obtained as:

$$f_{z_k}(z) = \frac{dF_{z_k}(z)}{dz} = -f_{\gamma_{k,RD}}\left(\frac{1}{z}\right)\left(-\frac{1}{z^2}\right) = \frac{1}{z^2}f_{\gamma_{k,RD}}\left(\frac{1}{z}\right). \tag{5.20}$$

Introducing the transformation $y_k = 1 + \rho z_k$, the PDF of the random variable y_k, representing the denominator in the relation (5.18), can be found. Thus, using the already derived PDF of $\gamma_{k,RD}$ given in (5.15), the PDF of the random variable y_k becomes:

$$f_{y_k}(y_k) = \rho\frac{U_{\{y_k-1\}}}{(y_k - 1)^2}\sum_{i=0}^{M-k}\lambda_{RD}\delta_i e^{-\varepsilon_i\lambda_{RD}\frac{\rho}{y_k-1}}, \tag{5.21}$$

where $U_{\{y_k-1\}}$ is the unit step function, given with:

$$U_{\{y_k-1\}} = \begin{cases} 1, & \text{for } y_k \geq 1 \\ 0, & \text{for } y_k < 1 \end{cases}. \tag{5.22}$$

Now, the variable $\gamma_{k,end}$ can be represented as a ratio of two random variables $(\gamma_{k,end} = \gamma_{k,SR}/y_k)$ whose PDFs are known, so that the PDF of the end-to-end SNR for the case of BTW SCP mapping could be defined as:

$$f_{\gamma_k,end}^{BTW}(x) = \int_0^\infty y_k f_{\gamma_{SR},y_k}(xy_k, y_k)dy_k, \qquad (5.23)$$

where $f_{\gamma_{SR},y_k}(xy_k, y_k)$ represents a joint PDF of the independent variables γ_{SR} and y_k, i.e.:

$$f_{\gamma_{SR},y}(xy_k, y_k) = f_{k,SR}^w(xy_k)f_{y_k}(y_k), \qquad (5.24)$$

and $f_{k,SR}^w(xy_k)$ is given with the relation (5.11). It follows that:

$$
\begin{aligned}
f_{\gamma_k,end}^{BTW}(x) &= \int_0^\infty y_k f_{k,SR}^w(xy_k)f_{y_k}(y_k)dy_k \\
&= \sum_{j=0}^{k-1}\sum_{i=0}^{M-k} \lambda_{SR}\lambda_{RD}\alpha_j\delta_i\rho\int_1^\infty \frac{y_k \exp\left(-\lambda_{SR}\beta_j xy_k - \lambda_{RD}\varepsilon_i\frac{\rho}{y_k-1}\right)}{(y_k-1)^2}dy_k.
\end{aligned}
$$
$$(5.25)$$

After introducing: $t = 1/(y_k - 1)$, the relation (5.25) becomes:

$$
\begin{aligned}
f_{\gamma_k,end}^{BTW}(x) &= \rho\lambda_{SR}\lambda_{RD}\sum_{j=0}^{k-1}\sum_{i=0}^{M-k} \exp(-\lambda_{SR}\beta_j x)\alpha_j\delta_i \\
&\times \int_0^\infty \left(1 + \frac{1}{t}\right)\exp\left(-\lambda_{RD}\varepsilon_i\rho t - \lambda_{SR}\beta_j x\frac{1}{t}\right)dt.
\end{aligned}
$$
$$(5.26)$$

The integral from the above relation can be separated in two standard integrals of the following form:

$$\int_0^\infty x^{\nu-1}e^{-(x+\mu^2/4x)}dx = 2\left(\frac{\mu}{2}\right)^2 K_{-\nu}(\mu), \quad \left(|\arg(\mu)| < \frac{\pi}{2};\ Re\{\mu^2\} > 0\right),$$
$$(5.27)$$

where $K_\nu(\cdot)$ is the second order modified Bessel function of the ν-th kind, characterized with $K_{-\nu}(\cdot) = K_\nu(\cdot)$. Subsequently, the expression for PDF of SNR of the k-th subcarrier at the output of the system with the BTW SCP

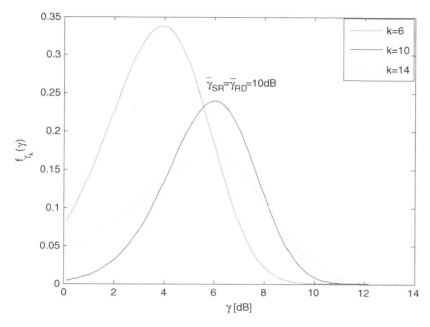

Fig. 5.2 PDF of the k-th subcarrier received SNR for the relay system with BTW SCP.

scheme, can be derived as:

$$f_{\gamma_k,end}^{BTW}(x) = \frac{2}{\bar{\gamma}_{SR}} \sum_{j=0}^{k-1} \sum_{i=0}^{M-k} \alpha_j \delta_i e^{-\beta_j \frac{x}{\bar{\gamma}_{SR}}} \left[\sqrt{\frac{\rho \beta_j x}{\varepsilon_i \bar{\gamma}_{SR} \bar{\gamma}_{RD}}} K_1 \left(2\sqrt{\frac{\rho \beta_j \varepsilon_i x}{\bar{\gamma}_{SR} \bar{\gamma}_{RD}}} \right) \right.$$
$$\left. + \frac{\rho}{\bar{\gamma}_{RD}} K_0 \left(2\sqrt{\frac{\rho \beta_j \varepsilon_i x}{\bar{\gamma}_{SR} \bar{\gamma}_{RD}}} \right) \right]. \tag{5.28}$$

Figure 5.2 shows PDF of SNR graphs for the k-th subcarrier ($k = 6, k = 10$, $k = 14$) at the receiving end of the OFDM AF FG relay system with the applied BTW SCP scheme. The illustrated examples are given for the case when the average SNR at the S-RS link is equal to 10dB, and it is identical with the average SNR at the RS-D link. It is assumed that the total number of subcarriers (or subcarrier groups) in the analyzed OFDM AF relay system is 16 ($M = 16$).

The presented graphs show that, for the BTW SCP scheme, there is no significant difference between the average SNRs of different subcarriers at the receiving end of the system. This means that BER values per subcarrier will also be close, as well as the system capacities per subcarrier. Further on, it can be concluded that the increase of the subcarrier order k causes at first the increase of the mean SNR value (for $k = 10$ the mean SNR value is higher than for $k = 6$), and then its decrease (for $k = 14$ the mean SNR value is lower than for $k = 10$). It can be expected that the relation between corresponding BERs per subcarrier, as well as between capacities per subcarrier, will follow the same tendency.

5.2.3   PDF of SNR for BTB SCP Scheme

PDF of the signal-to-noise ratio at the output of the OFDM AF FG relay system with BTB SCP scheme could be found applying the same algorithm as the one described for the BTW SCP scheme, with the exception of the fact that in this case RS sorts subcarriers on both S-RS and RS-D links in increasing order. This means that the PDF of the SNR for the k-th weakest subcarrier on the S-RS link could be presented with the relation (5.11). The same is valid for the PDF of the SNR for the k-th weakest subcarrier on the RS-D link, under the condition that $\bar{\gamma}_{SR}$ is replaced with $\bar{\gamma}_{RD}$, representing the average SNR for the RS-D link. Thus, the required PDF is obtained in the following form:

$$f_{\gamma_{k},end}^{BTB}(x) = \frac{2}{\bar{\gamma}_{SR}} \sum_{j=0}^{k-1} \sum_{i=0}^{k-1} \alpha_j \alpha_i e^{-\beta_j \frac{x}{\bar{\gamma}_{SR}}} \left[\sqrt{\frac{\rho \beta_j x}{\beta_i \bar{\gamma}_{SR} \bar{\gamma}_{RD}}} K_1 \left(2\sqrt{\frac{\rho \beta_j \beta_i x}{\bar{\gamma}_{SR} \bar{\gamma}_{RD}}} \right) \right.$$

$$\left. + \frac{\rho}{\bar{\gamma}_{RD}} K_0 \left(2\sqrt{\frac{\rho \beta_j \beta_i x}{\bar{\gamma}_{SR} \bar{\gamma}_{RD}}} \right) \right] \tag{5.29}$$

Figure 5.3 illustrates PDFs of SNR for the k-th subcarrier at the receiving end of the OFDM AF relay system with fixed gain and BTB subcarrier mapping. It is assumed that the average SNRs on S-RS and RS-D links are identical and equal to 10 dB, while the system is realized with the total of $M = 16$ subcarriers (or subcarrier groups). The presented results show that with the increase of subcarrier order k, the mean SNR value at the system output also increases. Thus, improvements of the achieved capacity can be expected, as well as a decrease of BER values per subcarrier. At the same time, it is evident that performances per subcarrier at the system receiving end depend on the subcarrier order, since for $k = 6$ the mean SNR value at D is

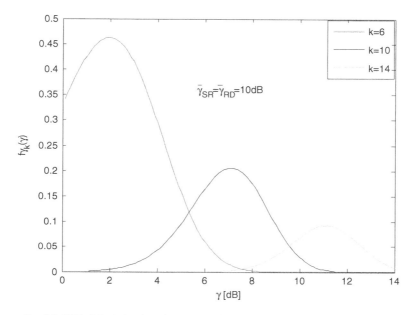

Fig. 5.3 PDF of the k-th subcarrier received SNR for the relay system with BTB SCP.

around 2 dB, while for $k = 14$ it comes to around 11 dB, (the average SNR per hop $= 10$ dB).

5.2.4 MGF of SNR for BTB SCP Scheme

MGF function is defined in a standard manner with:

$$\mathcal{M}_{\gamma_k, end}(s) = \mathbf{E}(e^{-s\gamma}) = \int_0^\infty f_{\gamma_k, end}(\gamma) e^{-s\gamma} d\gamma. \tag{5.30}$$

Introducing the previously derived expression (5.28) for the PDF of the k-th subcarrier SNR, the MGF of SNR for the case of BTW SCP can be obtained as:

$$\mathcal{M}_{\gamma_k, end}(s) = \sum_{j=0}^{k-1} \sum_{i=0}^{M-k} \frac{2\alpha_j \delta_i}{\bar{\gamma}_{SR}} \int_0^\infty e^{-x\left(s + \frac{\beta_j}{\bar{\gamma}_{SR}}\right)} \left[\sqrt{\frac{\rho \beta_j x}{\varepsilon_i \bar{\gamma}_{SR} \bar{\gamma}_{RD}}} K_1 \left(2\sqrt{\frac{\rho \beta_j \varepsilon_i x}{\bar{\gamma}_{SR} \bar{\gamma}_{RD}}} \right) \right.$$
$$\left. + \frac{\rho}{\bar{\gamma}_{RD}} K_0 \left(2\sqrt{\frac{\rho \beta_j \varepsilon_i x}{\bar{\gamma}_{SR} \bar{\gamma}_{RD}}} \right) \right] dx. \tag{5.31}$$

Separating the above given integral in two integrals, the MGF of the received SNR for the k-th subcarrier can be described with:

$$M_{\gamma_k,end}(s) = \frac{2}{\bar{\gamma}_{SR}} \sum_{j=0}^{k-1} \sum_{i=0}^{M-k} \alpha_j \delta_i (\mathcal{I}_1 + \mathcal{I}_2). \tag{5.32}$$

The integral $\mathcal{I}_1$ can be solved introducing the transformation $x = t^2$, what leads to the form that can be recognized as a standard one:

$$\int_0^\infty x^\mu e^{-\alpha x^2} K_\nu(\beta x) dx = \frac{1}{2} \alpha^{-\frac{1}{2}\mu} \beta^{-1} \Gamma\left(\frac{1+\mu+\nu}{2}\right)$$

$$\times \Gamma\left(\frac{1-\nu+\mu}{2}\right) e^{\frac{\beta^2}{8\alpha}} W_{-\frac{1}{2}\mu,\frac{1}{2}\nu}\left(\frac{\beta^2}{4\alpha}\right);$$

$$(Re(\mu) > |Re(\nu) - 1|). \tag{5.33}$$

Using the property of the Gamma function:

$$\Gamma(n+1) = n!, \ (n \text{ is an integer}), \tag{5.34}$$

the solution of the integral $\mathcal{I}_1$ could be written in the following form:

$$\mathcal{I}_1 = \frac{1}{2} \frac{1}{\varepsilon_i T_j(s)} e^{\frac{B_{j,i}}{T_j(s)}} W_{-1,\frac{1}{2}}\left(\frac{B_{j,i}}{T_j(s)}\right), \tag{5.35}$$

where:

$$T_j(s) = s + \beta_j/\bar{\gamma}_{SR} \tag{5.36}$$

and

$$B_{j,i} = \frac{\beta_j \varepsilon_i}{\bar{\gamma}_{SR} \bar{\gamma}_{RD}}. \tag{5.37}$$

The integral $\mathcal{I}_2$ can be solved if it is first transformed into the form given bellow:

$$\int_0^\infty e^{-\alpha x} K_{2\nu}(2\sqrt{\beta x}) dx = \frac{1}{2} \frac{1}{\sqrt{\alpha\beta}} e^{\frac{\beta}{2\alpha}} \Gamma(1+\nu)\Gamma(1-\nu) W_{-\frac{1}{2},\nu}\left(\frac{\beta}{\alpha}\right)$$

$$(Re(\alpha) > 0, \ |Re(\mu)| < 1), \tag{5.38}$$

which gives:

$$\mathcal{I}_2 = \frac{\rho}{2\bar{\gamma}_{RD}} \frac{1}{\sqrt{B_{j,i} T_j(s)}} e^{\frac{B_{j,i}}{2T_j(s)}} W_{-\frac{1}{2},0}\left(\frac{B_{j,i}}{T_j(s)}\right). \tag{5.39}$$

Thus, the final expression for the MGF of the end-to-end SNR for the k-th subcarrier, in the case of the OFDM AF FG relay system with BTW SCP, becomes:

$$\mathcal{M}_{\gamma_k,end}(s) = \frac{1}{\bar{\gamma}_{SR}} \sum_{j=0}^{k-1} \sum_{i=0}^{M-k} \alpha_j \delta_i \left[\frac{1}{T_j(s)\varepsilon_i} W_{-1,\frac{1}{2}} \left(\frac{B_{j,i}}{T_j(s)} \right) \right.$$

$$\left. + \frac{\rho}{\bar{\gamma}_{RD}\sqrt{B_{j,i}T_j(s)}} W_{-\frac{1}{2},0} \left(\frac{B_{j,i}}{T_j(s)} \right) \right]. \tag{5.40}$$

The obtained MGF expression could also be given in terms of exponential integral function, which is more suitable for further mathematical transformations than the Whittaker function present in (5.40). So, the following relations between the Whittaker and exponential integral function could be applied:

$$W_{-\frac{1}{2},0}(z) = e^{\frac{z}{2}} z^{\frac{1}{2}} E_1(z), \tag{5.41}$$

$$W_{-1,\frac{1}{2}}(z) = e^{\frac{z}{2}} (e^{-z} - z E_1(z)). \tag{5.42}$$

leading toward the following expression for the MGF of the end-to-end SNR:

$$\mathcal{M}_{\gamma_k,end}(s) = \frac{1}{\bar{\gamma}_{SR}} \sum_{j=0}^{k-1} \sum_{i=0}^{M-k} \frac{\alpha_j \delta_i}{T_j(s)}$$

$$\times \left[\frac{1}{\varepsilon_i} + e^{\frac{\rho B_{j,i}}{T_j(s)}} E_1 \left(\frac{\rho B_{j,i}}{T_j(s)} \right) \left(\frac{\rho}{\bar{\gamma}_{RD}} - \frac{\rho B_{j,i}}{\varepsilon_i T_j(s)} \right) \right]. \tag{5.43}$$

5.2.5 MGF of SNR for BTB SCP Scheme

Introducing the expression (5.29) for the PDF of SNR for the BTB SCP scheme into the relation (5.30), and using the same procedure applied in derivation of the MGF of SNR for the BTW SCP, the following expression for the MGF of the end-to-end k-th subcarrier SNR for the BTB SCP scheme can be obtained:

$$\mathcal{M}_{\gamma_k,end}(s) = \frac{1}{\bar{\gamma}_{SR}} \sum_{j=0}^{k-1} \sum_{i=0}^{k-1} \alpha_j \delta_i \left[\frac{1}{T_j(s)\beta_i} W_{-1,\frac{1}{2}} \left(\frac{A_{j,i}}{T_j(s)} \right) \right.$$

$$\left. + \frac{\rho}{\bar{\gamma}_{RD}\sqrt{A_{j,i}T_j(s)}} W_{-\frac{1}{2},0} \left(\frac{A_{j,i}}{T_j(s)} \right) \right], \tag{5.44}$$

where:

$$A_{j,i} = \beta_j \beta_i / \bar{\gamma}_{SR} \bar{\gamma}_{RD}. \tag{5.45}$$

Expressing in terms of the exponential integral function, the above relation becomes:

$$\mathcal{M}_{\gamma_k,end}(s) = \frac{1}{\bar{\gamma}_{SR}} \sum_{j=0}^{k-1} \sum_{i=0}^{k-1} \frac{\alpha_j \alpha_i}{T_j(s)}$$

$$\times \left[\frac{1}{\beta_i} + e^{\frac{\rho A_{j,i}}{T_j(s)}} E_1 \left(\frac{\rho A_{j,i}}{T_j(s)} \right) \left(\frac{\rho}{\bar{\gamma}_{RD}} - \frac{\rho A_{j,i}}{\beta_i T_j(s)} \right) \right]. \tag{5.46}$$

The complete set of the presented expressions for the PDFs and the MGFs, describing statistics of the received SNR for the two considered subcarrier mapping schemes, is essential for further performance analyses of multi-carrier OFDM AF relay systems with fixed gain.

5.3   BER Performance of OFDM AF FG Relay Systems with SCP

When deciding about incorporation of BTW and BTB subcarrier permutation schemes, it is necessary to have full knowledge about their impact on the performance of OFDM AF FG relay systems. Since performances like BER and capacity depend on the data symbol modulation, appropriate analyses should focus on both non-coherent (DPSK) and coherent (BPSK, m-QAM) modulation methods.

5.3.1   BER of DPSK Modulated OFDM AF FG Relay Systems with SCP

Using the derived expressions for the MGF of the end-to-end SNR, as well as the MGF approach in BER determination [82], it is possible to formulate an adequate BER determination algorithm for different input data modulation methods and the transmission over wireless fading channels. Thus, when for example DPSK modulation is applied, BER on the k-th subcarrier at the system receiving end is given with:

$$P_{b,k} = 0,5 \mathcal{M}_{\gamma_k,end}(1), \tag{5.47}$$

In order to derive BER relations for the considered OFDM AF FG relay system with BTW SCP, or BTB SCP, it is sufficient to introduce appropriate MGF expressions. Subsequently, the average BER at the receiving end of the analyzed OFDM relay system can be obtained by averaging over the set of BERs obtained for each subcarrier:

$$P_b = \frac{1}{M} \sum_{k=1}^{M} P_{b,k}. \tag{5.48}$$

Since the expression for the MGF of the received SNR, for the both subcarrier permutation schemes, includes the exponential integral function, MGF approach for BER determination becomes very complex when coherent modulation methods are considered. That is why, a different approach, based on the knowledge of probability density functions, will be applied for the systems where m-PSK or m-QAM are implemented.

5.3.2 BER of BPSK Modulated OFDM AF FG Relay Systems with SCP

Applying the mentioned PDF approach in analyzing BER performance of a system with coherent modulation and transmission over wireless fading channel, BER on the k-th subcarrier at the receiving end of the considered relay system can be defined as:

$$P_{b,k} = \int_0^\infty P_{b|\gamma_k} f_{\gamma_k,end}(\gamma) d\gamma, \tag{5.49}$$

where $P_{b|\gamma_k}$ denotes the conditional bit error rate, which depends on the modulation method applied, while $f_{\gamma_k,end}$ is the PDF of the end-to-end SNR for the k-th subcarrier. The conditional bit error rate $P_{b|\gamma_k}$, for input data symbols being BPSK modulated, is given with [102]:

$$P_{b|\gamma_k} = Q(\sqrt{2\gamma}). \tag{5.50}$$

In order to solve the integral in the relation (5.49), the following approximation of the complementary error function can be used [111]:

$$erfc(x) \cong \frac{1}{6} e^{-x^2} + \frac{1}{2} e^{-4x^2/3}. \tag{5.51}$$

Thus, using the above given approximation, the probability of error per bit for the k-th subcarrier at the system receiving end, becomes:

$$P_{b,k} = \int_0^\infty \left(\frac{1}{6} e^{-\gamma} + \frac{1}{2} e^{-4\gamma/3} \right) f_{\gamma_k,end}(\gamma) d\gamma. \tag{5.52}$$

For the BTW SCP scheme, as well as for BTB SCP, the integral in (5.52) can be transformed into the standard ones described with the relations (5.33) and (5.38). That is how the BER expression for the k-th subcarrier is obtained, in the form including Whittaker function. Applying the equations given with (5.41) and (5.42), the derived $P_{b,k}$ relation can be further on modified by introducing the exponential integral function. So, for the BTW SCP scheme, BER for the k-th subcarrier at the receiving end of the OFDM AF FG relay system, and the BPSK modulation, is obtained as:

$$
P_{b,k} = \frac{1}{2\bar{\gamma}_{SR}} \sum_{j=0}^{k-1} \sum_{i=0}^{M-k} \alpha_j \delta_i \left[\frac{1}{2\varepsilon_i} \left(\frac{1}{3T_j(1)} + \frac{1}{T_j(4/3)} \right) \right.
$$
$$
+ \frac{1}{6} \frac{e^{\frac{\rho B_{j,i}}{T_j(1)}}}{T_j(1)} E_1 \left(\frac{\rho B_{j,i}}{T_j(1)} \right) \left(\frac{\rho}{\bar{\gamma}_{RD}} - \frac{\rho B_{j,i}}{2\varepsilon_i T_j(1)} \right)
$$
$$
\left. + \frac{1}{2} \frac{e^{\frac{\rho B_{j,i}}{T_j(4/3)}}}{T_j(4/3)} E_1 \left(\frac{\rho B_{j,i}}{T_j(4/3)} \right) \left(\frac{\rho}{\bar{\gamma}_{RD}} - \frac{\rho B_{j,i}}{2\varepsilon_i T_j(4/3)} \right) \right], \quad (5.53)
$$

where:

$$
T_j(1) = 1 + \beta_j / \bar{\gamma}_{SR}, \quad (5.54)
$$
$$
T_j(4/3) = 4/3 + \beta_j / \bar{\gamma}_{SR}. \quad (5.55)
$$

Under the same set of conditions, BER for the k-th subcarrier at the receiving end of the OFDM AF FG relay system with BTB SCP becomes:

$$
P_{b,k} = \frac{1}{2\bar{\gamma}_{SR}} \sum_{j=0}^{k-1} \sum_{i=0}^{k-1} \alpha_j \beta_i \left[\frac{1}{2\beta_i} \left(\frac{1}{3T_j(1)} + \frac{1}{T_j(4/3)} \right) \right.
$$
$$
+ \frac{1}{6} \frac{e^{\frac{\rho A_{j,i}}{T_j(1)}}}{T_j(1)} E_1 \left(\frac{\rho A_{j,i}}{T_j(1)} \right) \left(\frac{\rho}{\bar{\gamma}_{RD}} - \frac{\rho A_{j,i}}{2\beta_i T_j(1)} \right)
$$
$$
\left. + \frac{1}{2} \frac{e^{\frac{\rho A_{j,i}}{T_j(4/3)}}}{T_j(4/3)} E_1 \left(\frac{\rho A_{j,i}}{T_j(4/3)} \right) \left(\frac{\rho}{\bar{\gamma}_{RD}} - \frac{\rho A_{j,i}}{2\beta_i T_j(4/3)} \right) \right]. \quad (5.56)
$$

For the both subcarrier permutation schemes, the total average BER value is defined through averaging over the set of BERs obtained for each subcarrier present at the receiving end of the system:

$$P_b = \frac{1}{M} \sum_{k=1}^{M} P_{b,k}. \tag{5.57}$$

5.3.3 BER of m-QAM Modulated OFDM AF FG Relay Systems with SCP

The previously described analytical approach for BER determination in the case of OFDM AF FG relay system with BPSK, can be further applied for any m-QAM modulation. It is just necessary to take into account the fact that the conditional bit error rate for m-QAM modulation has the following form [44]:

$$P_{b|\gamma_k} = \sum_{n=1}^{\sqrt{m}-1} \mathcal{A}_n Q(\sqrt{\mathcal{B}_n \gamma}). \tag{5.58}$$

where $\mathcal{A}_i$ and $\mathcal{B}_i$ are coefficients depending on the order m of the QAM modulation. For example, for 4-QAM modulation: $\mathcal{A}_i = 1$ and $\mathcal{B}_i = 1$, which gives:

$$P_{b|\gamma_k} = Q(\sqrt{\gamma}), \tag{5.59}$$

while for the 16-QAM modulation this conditional bit error rate becomes [44]:

$$P_{b|\gamma_k} = \frac{3}{4} Q\left(\sqrt{\frac{1}{5}\gamma}\right) + \frac{1}{2} Q\left(\sqrt{\frac{9}{5}\gamma}\right) - \frac{1}{4} Q(\sqrt{5\gamma}). \tag{5.60}$$

Introducing the relation (5.58) into the relation (5.49), and applying the approximation of the complementary error function (5.51), the required BER expression for the m-QAM OFDM AF FG relay system with SCP is derived:

$$P_{b,k} = \sum_{n=1}^{\sqrt{m}-1} \mathcal{A}_n \int_0^{\infty} \left(\frac{1}{6}e^{-\gamma \mathcal{B}_n/2} + \frac{1}{2}e^{-2\gamma \mathcal{B}_n/3}\right) f_{\gamma_k,end}(\gamma) d\gamma. \tag{5.61}$$

When the BTW subcarrier permutation scheme is implemented, BER for the k-th subcarrier at the system receiving end is calculated by introducing the relation (5.28) for the PDF of SNR into the relation (5.61). Then, transforming it into the standard integrals given with (5.33) and (5.38), and using the equations given with (5.41) and (5.42), the following form can be obtained:

$$
P_{b,k} = \frac{1}{2\bar{\gamma}_{SR}} \sum_{n=1}^{\sqrt{m}-1} \mathcal{A}_n \sum_{j=0}^{k-1} \sum_{i=0}^{M-k} \alpha_j \delta_i \left[\frac{1}{2\varepsilon_i} \left(\frac{1}{3\mathcal{T}_{n,j}(1/2)} + \frac{1}{\mathcal{T}_{n,j}(2/3)} \right) \right.
$$

$$
+ \frac{1}{6} \frac{e^{\frac{\rho B_{j,i}}{\mathcal{T}_{n,j}(1/2)}}}{\mathcal{T}_{n,j}(1/2)} E_1 \left(\frac{\rho B_{j,i}}{\mathcal{T}_{n,j}(1/2)} \right) \left(\frac{\rho}{\bar{\gamma}_{RD}} - \frac{\rho B_{j,i}}{2\beta_i \mathcal{T}_{n,j}(1/2)} \right)
$$

$$
\left. + \frac{1}{2} \frac{e^{\frac{\rho B_{j,i}}{\mathcal{T}_{n,j}(2/3)}}}{\mathcal{T}_{n,j}(2/3)} E_1 \left(\frac{\rho B_{j,i}}{\mathcal{T}_{n,j}(2/3)} \right) \left(\frac{\rho}{\bar{\gamma}_{RD}} - \frac{\rho B_{j,i}}{2\beta_i \mathcal{T}_{n,j}(2/3)} \right) \right], \quad (5.62)
$$

where:

$$
\mathcal{T}_{n,j}(x) = x\mathcal{B}_n + \beta_j/\bar{\gamma}_{SR}. \quad (5.63)
$$

For the BTB SCP scheme, BER for the k-th subcarrier at the receiving end of the analyzed relay system with m-QAM is described with:

$$
P_{b,k} = \frac{1}{2\bar{\gamma}_{SR}} \sum_{n=1}^{\sqrt{m}-1} \mathcal{A}_n \sum_{j=0}^{k-1} \sum_{i=0}^{k-1} \alpha_j \beta_i \left[\frac{1}{2\beta_i} \left(\frac{1}{3\mathcal{T}_{n,j}(1/2)} + \frac{1}{\mathcal{T}_{n,j}(2/3)} \right) \right.
$$

$$
+ \frac{1}{6} \frac{e^{\frac{\rho A_{j,i}}{\mathcal{T}_{n,j}(1/2)}}}{\mathcal{T}_{n,j}(1/2)} E_1 \left(\frac{\rho A_{j,i}}{\mathcal{T}_{n,j}(1/2)} \right) \left(\frac{\rho}{\bar{\gamma}_{RD}} - \frac{\rho A_{j,i}}{2\beta_i \mathcal{T}_{n,j}(1/2)} \right)
$$

$$
\left. + \frac{1}{2} \frac{e^{\frac{\rho A_{j,i}}{\mathcal{T}_{n,j}(2/3)}}}{\mathcal{T}_{n,j}(2/3)} E_1 \left(\frac{\rho A_{j,i}}{\mathcal{T}_{n,j}(2/3)} \right) \left(\frac{\rho}{\bar{\gamma}_{RD}} - \frac{\rho A_{j,i}}{2\beta_i \mathcal{T}_{n,j}(2/3)} \right) \right]. \quad (5.64)
$$

5.4　Ergodic Capacity of OFDM AF FG Relay Systems with SCP

The previously derived relations for the PDFs of the received signal-to-noise ratio for the BTW SCP and BTP SCP schemes represent the basis for analyzing the achievable ergodic capacity of the considered OFDM AF FG relay systems with subcarrier permutation. Actually, the ergodic capacity of the k-th subcarrier at the system receiving end, with any of the two subcarrier permutation schemes, can be defined with:

$$
C_k = \frac{1}{2}\mathbf{E}\left(\log_2(1+\gamma)\right) = \frac{1}{2}\int_0^\infty \log_2(1+\gamma) f_{\gamma_{k,end}}(\gamma)d\gamma, \quad (5.65)
$$

where $f_{\gamma_{k,end}} \in \left\{ f_{\gamma_{k,end}}^{BTW}, f_{\gamma_{k,end}}^{BTB} \right\}$, and the coefficient 1/2 is introduced due to realization of data transmission over the two time slots. The integral in the relation (5.65) can not be directly solved in the closed form, due to the fact that PDF of SNR expressions for BTW SCP and BTB SCP include the second order modified Bessel functions. However, taking into account the concave form of the logarithmic function, and the Jensen's inequality which establishes the following relation between a random variable x and a concave function φ:

$$\mathbf{E}(\varphi(x)) \leq \varphi(\mathbf{E}(x)), \tag{5.66}$$

the upper bound of the ergodic capacity for the k-th subcarrier at the system receiving end can be found. Thus, applying the Jensen's inequality on the relation for the ergodic capacity of the k-th subcarrier at the system receiving end, it is obtained that:

$$C_k \leq \frac{1}{2}\log_2(1 + \mathbf{E}(\gamma)). \tag{5.67}$$

The expectation of the signal-to-noise ratio for the k-th received subcarrier is given with:

$$\mathbf{E}(\gamma) = \int_0^\infty \gamma f_{\gamma_{k,end}}(\gamma)d\gamma, \tag{5.68}$$

and it can be calculated using the integral described with the relation (5.33). When the BTW SCP is implemented, $\mathbf{E}(\gamma)$ has the following form:

$$\mathbf{E}(\gamma) = \sum_{j=0}^{k-1}\sum_{i=0}^{M-k} \frac{\alpha_j \delta_i}{\beta_j^2 \sqrt{\rho B_{j,i}} e^{\frac{\rho B_{j,i}\bar{\gamma}_{SR}}{2\beta_j}}} \left[2\sqrt{\frac{\rho\beta_j}{\varepsilon_i}\frac{\bar{\gamma}_{SR}}{\bar{\gamma}_{RD}}} \, W_{-2,\frac{1}{2}}\left(\frac{\rho B_{j,i}\bar{\gamma}_{SR}}{\beta_j}\right) \right.$$
$$\left. + \frac{\rho\sqrt{\beta_j}\bar{\gamma}_{SR}}{\bar{\gamma}_{RD}} \, W_{-\frac{3}{2},0}\left(\frac{\rho B_{j,i}\bar{\gamma}_{SR}}{\beta_j}\right) \right], \tag{5.69}$$

while for the BTB SCP, $\mathbf{E}(\gamma)$ becomes:

$$\mathbf{E}(\gamma) = \sum_{j=0}^{k-1}\sum_{i=0}^{k-1} \frac{\alpha_j \alpha_i}{\beta_j^2 \sqrt{\rho A_{j,i}} e^{\frac{\rho A_{j,i}\bar{\gamma}_{SR}}{2\beta_j}}} \left[2\sqrt{\frac{\rho\beta_j}{\beta_i}\frac{\bar{\gamma}_{SR}}{\bar{\gamma}_{RD}}} \, W_{-2,\frac{1}{2}}\left(\frac{\rho A_{j,i}\bar{\gamma}_{SR}}{\beta_j}\right) \right.$$
$$\left. + \frac{\rho\sqrt{\beta_j}\bar{\gamma}_{SR}}{\bar{\gamma}_{RD}} \, W_{-\frac{3}{2},0}\left(\frac{\rho A_{j,i}\bar{\gamma}_{SR}}{\beta_j}\right) \right]. \tag{5.70}$$

The overall ergodic capacity of the OFDM AF FG system with SCP is calculated by summing up ergodic capacities determined for each subcarrier, or subcarrier group:

$$C = \sum_{k=1}^{M} C_k. \tag{5.71}$$

5.5   Performance Analysis of OFDM AF FG Relay Systems with SCP

In this section, performances of OFDM AF FG relay systems with SCP, obtained analytically, and verified through simulations, are analyzed. Unless otherwise stated, the ideal time and frequency system synchronization is assumed. The incorporated OFDM system has M subcarriers, or M subcarrier groups in a real scenario, with mutually uncorrelated transfer functions. The presented analyses can be considered as valid regardless the exact number of subcarriers, as long as they are uncorrelated. Thus, for the reason of efficiency, M is taken to be 16. At the same time, all given analytical and simulation models start from the assumption that the average energy per symbol transmitted by RS is equal with the average energy per symbol transmitted by S, i.e., $\epsilon_R = \epsilon_S$, and the noise variances at RS and D are also identical, i.e., $\mathcal{N}_{01} = \mathcal{N}_{02} = \mathcal{N}_0$.

The fixed gain G applied at the relay station could be defined using the assumption that RS has the knowledge of the S-RS channel state:

$$G^2 = \mathbf{E}\left[\frac{\epsilon_R}{\epsilon_S|H_{1,k}|^2 + \mathcal{N}_0}\right], \tag{5.72}$$

which for the Rayleigh fading distribution gives:

$$G^2 = \frac{\epsilon_R}{\epsilon_S E\left[|H_{1,k}|^2\right]} e^{1/\bar{\gamma}_{SR}} E_1\left(\frac{1}{\bar{\gamma}_{SR}}\right). \tag{5.73}$$

Thus, the previously established relation between the coefficient ρ and G, $\rho = \epsilon_R/(G^2 \mathcal{N}_0)$, becomes :

$$\rho = \frac{\bar{\gamma}_{SR}}{e^{1/\bar{\gamma}_{SR}} E_1(1/\bar{\gamma}_{SR})}. \tag{5.74}$$

Simulation results presented in the following subparagraphs are obtained using Monte Carlo method, with the system OFDM part being simulated in the frequency domain. This approach can be considered as valid since the ideal time and frequency synchronization is assumed. Channel with Rayleigh fading is modeled in a such way that the subcarrier (or group of subcarriers) transfer function is generated as a complex sum of the two independent Gaussian variables with the zero mean and the variance equal to 1/2, which means that the average power of each subcarrier is taken to be equal to 1. It is obtained that the subcarriers generated in this way are not correlated, and the transmission of ten OFDM symbols is assumed for each simulated subcarrier channel. Binary symbols are generated as a random sequence of 1 s and 0 s, having a uniform distribution, with 0 s being then converted into -1 s, so that each symbol carries the same energy. At both RS and D, white Gaussian noise is added, with variances $\mathcal{N}_{01}$ and $\mathcal{N}_{02}$, respectively. These variances depend on the given average SNR ratio on S-RS and RS-D links. RS performs subcarrier permutation in accordance to instantaneous SNR values of subcarriers on S-RS and RS-D links. The SCP function is assumed to be known to terminal D, which, after the equalization, reorders subcarriers and performs demodulation. The obtained information bits are then compared with the ones sent by S, the errors are summed for each iteration, and at the end the average error value for the given average SNR values on hops is calculated. More details on simulation of system ergodic capacity are given in the paragraph dealing with this topic.

The above described parameters are applied in simulations presented throughout the following analyses.

5.5.1 BER Performance Analysis of DPSK Modulated OFDM AF FG Relay Systems with SCP

Applying the MGF approach and using the obtained MGFs of SNR for the OFDM AF FG relay system with DPSK symbol modulation and SCP, BER values showed in Figure 5.4 are obtained. For the sake of comparison, the figure also includes the values for the analyzed system without (w/o) SCP is incorporated. These BER values are calculated using the adequate relation for the MGF of the end-to-end SNR in the OFDM AF FG system without SCP [62]:

$$\mathcal{M}_{\gamma_k,end}(s) = \frac{1}{\bar{\gamma}_{SR}s + 1} + \frac{\rho\bar{\gamma}_{SR}se^{\rho(\bar{\gamma}_{SR}s+1)/\bar{\gamma}_{RD}}}{\bar{\gamma}_{RD}(\bar{\gamma}_{SR}s + 1)^2} E_1\left(\frac{\rho}{\bar{\gamma}_{RD}(\bar{\gamma}_{SR}s + 1)}\right).$$

$$(5.75)$$

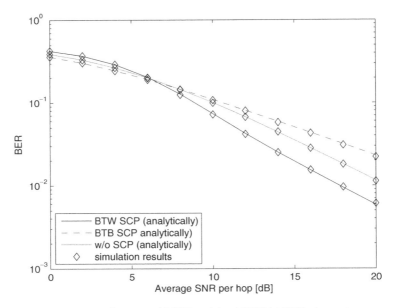

Fig. 5.4 BER performance of DPSK modulated OFDM AF FG relay system.

It is evident that the simulation results match with the ones obtained analytically, proving thus the validity of the analytical approach described in this chapter. As expected, for small SNRs, it is the BTB SCP scheme which has the best performance. More precisely, in the presented scenario when the average SNR of the S-RS link is equal to the average SNR of the RS-D link, the BTB SCP scheme gives the best BER values for SNRs bellow 6,5 dB. For SNRs above 6,5 dB, it is the BTW SCP scheme that outperforms the other options, with this advantage being preserved as SNR further increases.

5.5.2 BER Performance Analysis of BPSK Modulated OFDM AF FG Relay Systems with SCP

Using the PDF approach for BER calculations and approximation of the complementary error function, BER expressions of the BPSK modulated OFDM AF FG relay systems with BTW SCP and BTB SCP were derived. Figure 5.5 shows the adequate BER graphs obtained analytically, as well as by simulations.

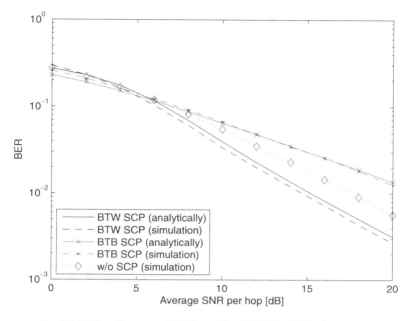

Fig. 5.5 BER performance of BPSK modulated OFDM AF FG relay system.

Again, the validity of the applied analytical model is clear, despite the introduced approximation. The difference between the BER values obtained analytically and by simulations is less than 0,5 dB over the complete SNR range, even for the worst case when the BTW SCP scheme is implemented. Generally, the results shown in Figure 5.5 illustrate that for the BPSK data symbol modulation, in the region of low SNR per hop, the best solution from the point of BER performance, is the BTB SCP scheme. On the other side, for the higher SNRs, the BTW SCP scheme becomes a better choice.

5.5.3 BER Performance Analysis of 4-QAM Modulated OFDM AF FG Relay Systems with SCP

As an example in BER performance analysis of m-QAM modulated OFDM AF FG relay systems with SCP, 4-QAM modulation can be chosen. The graphs illustrating the BER values obtained applying the previously presented analytical approach, as well as the ones obtained by simulations, are given in

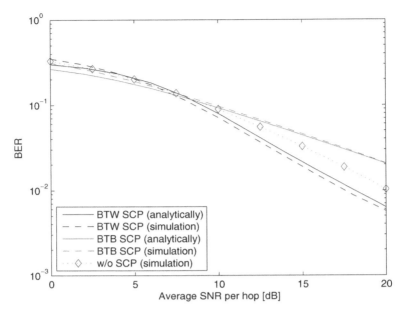

Fig. 5.6 BER performance of 4-QAM modulated OFDM AF FG relay system.

Figure 5.6. The propagation scenario considered assumes that the average SNRs on both, S-RS and RS-D, links are equal.

Comparisons of the presented graphs show that the analytically calculated BER values for 4-QAM approximate well the real BER values in the whole range of the average SNRs per hop considered. Thus, full analyses of BER performance for the m-QAM modulated OFDM AF FG relay system of interest, is enabled. In the particular situation presented in Figure 5.6, it can be noticed that the analytical and simulation BER values differ less than 0,5 dB for the BTW SCP scheme, while for the BTB SCP scheme this difference becomes even smaller. On the other side, the BTW SCP scheme gives the best BER performance, for the values of the average SNR per hop above 6,5 dB and its SNR gain in comparison with the system without SCP is around 2 dB, when BER is equal to 10^{-2}.

The presented BER performance and their analyses for DPSK, BPSK and 4-QAM modulated OFDM AF FG relay system, clearly show the positive impact of subcarrier mapping applied at the relay station. It is evident that SCP generally enables significant BER performance improvements, what recommends it for the implementation in future WWAN networks. Further on, the presented analyses show that, in order to achieve the optimal BER performance, the relay station should apply one of the two considered subcarrier mapping scheme (BTB or BTW) depending on the average SNRs on the links involved.

5.5.4 Ergodic Capacity Analysis of OFDM AF FG Relay Systems with SCP

Using the analytical method described in paragraph 5.4, the other performance of interest — the upper bound of the achievable capacity can be analyzed. For the considered OFDM AF FG relay system with SCP and $M = 16$ subcarriers, the analytically calculated values, as well as the ones obtained by simulations, are illustrated in Figure 5.7. The presented graphs are given for different propagation conditions, modeled with the average SNR per hop being equal to 4 dB, 10 dB or 15 dB.

Simulations include generation of independent Rayleigh fading channel transfer functions, for each subcarrier on the S-RS link ($H_{1,k}^{(n)}$) and on the

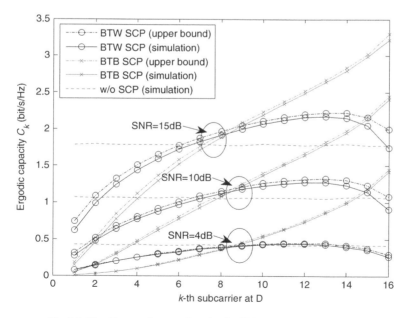

Fig. 5.7 Ergodic capacity per subcarrier for OFDM AF FG relay system.

RS-D link ($H_{2,k}^{(n)}$), and for each channel realization, where n denotes the n-th channel realization ($1 \leq n \leq n_{tot}$). The value of the ergodic capacity for the k-th subcarrier at the system receiving end, for the considered n-th simulation realization is defined as:

$$C_k^{(n)} = \frac{1}{2} \log_2 \left(1 + \frac{(G^{(n)})^2 |H_{1,k}^{(n)} H_{2,k}^{(n)} X_k^{(n)}|^2}{(G^{(n)})^2 |H_{2,k}^{(n)}|^2 \mathcal{N}_{0_1} + \mathcal{N}_{0_2}} \right). \tag{5.76}$$

The ergodic capacity for the k-th subcarrier at the receiving end of the considered relay system is then found by averaging over n_{tot} simulation repetitions:

$$C_k = \frac{1}{n_{tot}} \sum_{n=1}^{n_{tot}} C_k^{(n)}, \tag{5.77}$$

while the overall ergodic capacity of the whole relay system is given with: $C = \sum_{k=1}^{M} C_k$.

These values of ergodic capacities obtained by simulations are very close to the upper bounds calculated analytically, what is evident from Figure 5.7. There, it can be noticed that for the BTB SCP scheme the ergodic capacity increases as the subcarrier order k becomes higher, what can be considered as expected since the subcarrier SNR improves in the same manner. When the BTW SCP scheme is analyzed, the subcarrier with the highest SNR at the RS-D link, mapped with the subcarrier with the lowest SNR at the S-RS link, is denoted with $k = 1$, while $k = 16$ represents the opposite situation. Therefore, having the subcarrier channels with Rayleigh fading, the ergodic capacity at first increases with the increase of the subcarrier order k, and then the tendency changes in other direction. On the other side, when the relay system does not apply SCP, the ergodic capacity has the same value for each received subcarrier.

Table 5.1 contains analytical and simulation results for the average ergodic capacity per subcarrier for the systems implementing BTW SCP, BTB SCP, as well as the system without SCP in the scenario with equal average SNRs on both hops. The comparison of the analytically obtained results and simulation results for the OFDM AF FG relay system with BTB SCP shows that the obtained upper bound differs up to 3,3% from the simulation results, while in the case of the system implementing BTW SCP this difference goes up to 5%. The more important result which can be seen from the given table is that the highest ergodic capacity enhancement with BTB SCP scheme is achieved

Table 5.1. Average ergodic capacity per subcarrier at the receiving end of OFDM AF FG relay systems as function of average SNR per hop.

	SNR=4 dB		SNR=10 dB		SNR=15 dB	
	Simul.	Analyt.	Simul.	Analyt.	Simul.	Analyt.
BTW SCP [b/s/Hz]	0,322	0,335	0,979	1,027	1,741	1,82
BTB SCP [b/s/Hz]	0,496	0,512	1,144	1,176	1,848	1,898
No SCP [b/s/Hz]	0,416		1,06		1,785	

in the region of very small average SNR values on both hops, i.e., when propagation conditions on both hops are not favorable. Thus, for example, the system with BTB SCP has 19,2% higher capacity than the system without SCP for the average SNR values on both hops equal to 4 dB.

The given results also provide an insight in a decrease of ergodic capacity if the BTW SCP scheme is implemented, i.e., in the case when BER performance is of the primary interest. If simulation results for the ergodic capacity per subcarrier for the average SNR per hop value of 4 dB are considered, then it can be seen that the capacity decrease compared to the system with BTB SCP is equal to 35,3%, while it is equal to 22,9% in comparison to system without SCP. With the increase of the average SNR per values, the capacity lost due to implementation of BTW SCP scheme decreases.

The presented analyses clearly show that the incorporation of the BTB SCP scheme significantly improves the ergodic capacity performance of the OFDM AF FG relay systems. For further considerations in the sense of applications in future generation wireless networks, it is especially important to notice that the highest capacity improvements are achieved in the region of small average SNRs, considered critical from the point of high speed broadband wireless networks implementations.

6

Performance of OFDM AF VG Relay Systems with Subcarrier Permutation

In the previous chapter, the performance analyses of OFDM AF FG relay systems have shown that an adequate choice of subcarrier permutation scheme, enables significant performance enhancements in terms of both BER and the system capacity. In this part, similar analyses will be presented for OFDM AF VG relay systems, in order to examine the impact of subcarrier permutation scheme on their performance. Therefore, the closed form expressions for PDF and MGF of SNR at the system receiving end are derived at first. Then, it becomes possible to define the system BER values [32], as well as the upper bounds of the system capacity [31].

The approach applied in derivation of PDF of SNR and MGF expressions is partially different from the one applied for the OFDM AF system with fixed gain. This is due to the fact that, in addition to the statistics of ordered random variables, a relation existing between the system end-to-end SNR and the harmonic mean of two random variables is also included in the presented algorithm.

6.1 System Description

In analyzing this type of relay system the same communication scenario considered for the OFDM AF FG relay system is assumed. Thus, it is a dual-hop relay system with three terminals, where the complete communication between the terminal S and the terminal D is performed via the relay station (RS).

OFDM Based Relay Systems for Future Wireless Communications, 107–134.

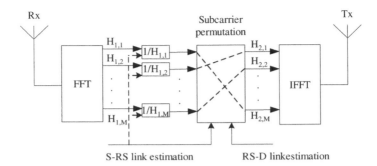

Fig. 6.1 Block diagram of OFDM AF relay station with variable gain and subcarrier permutation.

The required orthogonality among the S-RS link and the RS-D link is achieved by performing the communication process over two time slots. Further on, RS ideally knows the channel transfer function for each subcarrier on both hops, and the ideal time and frequency synchronizations between all terminals is assumed. In this relay system, after the OFDM demodulation in the FFT block (Figure 6.1), RS amplifies the i-th subcarrier signal with the factor G_i, which is inversely proportional to the channel transfer function for the i-th subcarrier on the S-RS link:

$$G_i = \frac{1}{H_{1,i}}, \quad 1 \le i \le M. \tag{6.1}$$

After that, and before the OFDM modulation in the IFFT block, the subcarrier permutation (SCP) is applied, depending on SNR values for each subcarrier on both hops. In order to perform the signal demodulation at the system receiving end, the terminal D has to know the subcarrier permutation scheme (modeled with $\upsilon(i)$) applied at RS.

For the system with VG, the signal received at the relay station RS, on the i-th subcarrier after the FFT block, is given with:

$$Y_{R,i} = X_i H_{1,i} + N_{1,i}, \quad 1 \le i \le M. \tag{6.2}$$

Assuming that the SCP function $\upsilon(i)$ performs mapping of the i-th subcarrier from the first hop to the k-th subcarrier on the second hop, the frequency domain representation of the signal received at D becomes:

$$
\begin{aligned}
Y_{D,k} &= G_i H_{2,k} Y_{R,\upsilon(i)} + N_{2,k} \\
&= G_i H_{2,k} H_{1,i} X_i + G_i H N_{1,i} + N_{2,k}, \quad 1 \le k \le M. \tag{6.3}
\end{aligned}
$$

while the k-th subcarrier signal-to-noise ratio (SNR) at the system receiving end is defined as:

$$\gamma_{k,end} = \frac{G_i^2 \mathbf{E}\{|X_{1,i}|^2\}|H_{1,i}|^2|H_{2,k}|^2}{\mathcal{N}_{02} + G_i^2|H_{2,k}|^2\mathcal{N}_{01}} = \frac{\frac{\mathbf{E}\{|X_{1,i}|^2\}|H_{1,i}|^2}{\mathcal{N}_{01}}\frac{|H_{2,k}|^2}{\mathcal{N}_{02}}}{\frac{|H_{2,k}|^2}{\mathcal{N}_{02}} + \frac{1}{G_i^2\mathcal{N}_{01}}} \tag{6.4}$$

Introducing the gain G_i, described with (6.1.1), into the relation given above, the final expression for the k-th subcarrier end-to-end SNR can be derived in the following form:

$$\gamma_{k,end} = \frac{\gamma_{i,SR}\gamma_{k,RD}}{\gamma_{i,SR} + \gamma_{k,RD}}. \tag{6.5}$$

As in the case of the OFDM AF relay system with fixed gain, it is assumed that the fadings among the subcarriers are independent and with Rayleigh distribution.

6.2 Statiscs of the End-to-End SNR

In order to analyze performance of communication systems characterized with communication channels with fading, it is necessary to define PDF (and CDF) of SNR at the system receiving end. Thus, it should be also done for the OFDM AF VG relay system in the assumed scenario with the Rayleigh fading on the links involved. Following the algorithm presented for the OFDM AF FG relay system, the statistics of the ordered random variables is used, with the difference that for the OFDM AF relay system with variable gain a relation between the harmonic mean of two random variables and the system end-to-end SNR should be first derived.

6.2.1 Harmonic Mean of Random Variables

The harmonic mean of two random variables a_1 and a_2 is defined as:

$$\mu_H(a_1, a_2) = \frac{2}{\frac{1}{a_1} + \frac{1}{a_2}} = \frac{2a_1 a_2}{a_1 + a_2} = a. \tag{6.6}$$

Comparing the relations (6.5) and (6.6), it becomes obvious that the k-th subcarrier received SNR for OFDM AF VG system could be expressed using the harmonic mean of the i-th subcarrier instantaneous SNR on the S-RS link and the k-th subcarrier instantaneous SNR on the RS-D link:

$$\mu_H(\gamma_{i,SR}, \gamma_{k,RD}) = \frac{2}{\frac{1}{\gamma_{i,SR}} + \frac{1}{\gamma_{k,RD}}} = \frac{2\gamma_{i,SR}\gamma_{k,RD}}{\gamma_{i,SR} + \gamma_{k,RD}} = 2\gamma_{k,end}. \tag{6.7}$$

Taking into account the above relation between the k-th subcarrier end-to-end SNR and the harmonic mean of the two instantaneous SNRs, it becomes possible to establish an adequate relation between their corresponding PDF functions:

$$f_{\gamma_k,end}(x) = 2f_a(2x). \tag{6.8}$$

Applying the approach used for the determination of the PDF for the harmonic mean of two exponentially distributed random variables [72], what corresponds with the Rayleigh fading subcarrier distribution, and using the relation (6.8), it becomes possible to derive the PDF of the end-to-end SNR for BTW SCP and BTB SCP implemented in OFDM AF VG relay systems.

6.2.2   PDF of SNR for BTW SCP Scheme

In order to define the end-to-end PDF of SNR for OFDM AF VG relay systems with BTW SCP scheme, the relation for the harmonic mean of two random variables (6.6) can be used, assuming that the variable a_1 represents the k-th smallest variable from the set of M i.i.d. exponentially distributed variables, while a_2 is the k-th biggest variable from the set having also M i.i.d. exponentially distributed variables. If the narrowband fading on both S-RS and RS-D links has Rayleigh statistics, then the variable a_1 actually correspondents to the k-th subcarrier SNR on the S-RS link, obtained after sorting at RS the subcarriers from the first hop in increasing order in regard to their SNRs. At the same time, the variable a_2 corresponds to SNR of the k-th best subcarrier on the RS-D link, i.e., SNR of the subcarrier which has the k-th highest SNR from the set of M subcarriers. Using these assumptions, the PDF of the variable a_1 is given with the relation (5.2.4), while the relation (5.2.8) describes the PDF of the variable a_2. New random variables are introduced:

$$z_1 = \frac{1}{a_1} \quad \text{and} \quad z_2 = \frac{1}{a_2}. \tag{6.9}$$

The PDF of the variable z_1 can be expressed with:

$$f_{z_1}(z) = \frac{1}{z^2} f^w_{\gamma_k,end}\left(\frac{1}{z}\right) = \frac{\lambda_{SR}}{z^2} \sum_{j=0}^{k-1} \lambda_{SR}\alpha_j e^{-\beta_j \lambda_{SR}\frac{1}{z}}, \qquad (6.10)$$

and the PDF of the variable z_2 as:

$$f_{z_2}(z) = \frac{1}{z^2} f^s_{\gamma_k,end}\left(\frac{1}{z}\right) = \frac{\lambda_{RD}}{z^2} \sum_{i=0}^{M-k} \lambda_{RD}\delta_i e^{-\varepsilon_i \lambda_{RD}\frac{1}{z}}. \qquad (6.11)$$

The MGF function of the variable z_1 is described with the following expression:

$$\mathcal{M}_{z_1}(s) = \mathbf{E}\left(e^{-sz}\right) = \lambda_{SR} \sum_{j=0}^{k-1} \alpha_j \int_0^\infty \frac{1}{z^2} \exp\left(-sz - \lambda_{SR}\beta_j \frac{1}{z}\right) dz. \quad (6.12)$$

Using the integral:

$$\int_0^\infty x^{v-1} \exp\left(-\frac{\beta}{x} - \gamma x\right) dx = 2\left(\frac{\beta}{\gamma}\right)^{0.5v} K_v(2\sqrt{\beta\gamma}),$$

$$(\text{Re}(\beta) > 0, \text{Re}(v) > 0), \qquad (6.13)$$

the MGF function of the variable z_1 becomes:

$$\mathcal{M}_{z_1}(s) = 2\lambda_{SR} \sum_{j=0}^{k-1} \alpha_j \sqrt{\frac{s\lambda_{SR}}{\beta_j}} K_1(2\sqrt{s\lambda_{SR}\beta_j}). \qquad (6.14)$$

Applying the same approach, the MGF function of the variable z_2 can be derived as:

$$\mathcal{M}_{z_2}(s) = 2\lambda_{RD} \sum_{i=0}^{M-k} \delta_i \sqrt{\frac{s\lambda_{RD}}{\varepsilon_i}} K_1(2\sqrt{s\lambda_{RD}\beta_i}). \qquad (6.15)$$

Further on, a new variable z, which is an inverse value of the variable a representing the harmonic mean of a_1 and a_2, is introduced:

$$z = \frac{1}{a} = \frac{1}{\mu_H(a_1, a_2)} = \frac{1}{2}\left(\frac{1}{a_1} + \frac{1}{a_2}\right). \qquad (6.16)$$

Knowing the properties of MGF functions, the MGF of the new variable z can be determined as:

$$\mathcal{M}_z(s) = \frac{1}{2}\mathcal{M}_{z_1}(s)\mathcal{M}_{z_2}(s), \qquad (6.17)$$

i.e.,:

$$\mathcal{M}_z(s) = 2\sum_{j=0}^{k-1}\sum_{i=0}^{M-k}\alpha_j\delta_i s\sqrt{\frac{\lambda_{SR}\lambda_{Rd}}{\beta_j\varepsilon_i}}K_1(2\sqrt{s\lambda_{SR}\beta_j})K_1(2\sqrt{s\lambda_{RD}\varepsilon_i}). \quad (6.18)$$

The CDF function of the variable a can be defined using the CDF function of the variable z:

$$F_a(A) = P(a < A) = P\left(\frac{1}{a} > \frac{1}{A}\right) = P\left(z > \frac{1}{A}\right) = 1 - P\left(z < \frac{1}{A}\right), \quad (6.19)$$

which gives:

$$F_a(Y) = 1 - F_z\left(\frac{1}{A}\right). \quad (6.20)$$

In order to determine the CDF function of the variable z, the previously defined MGF function, as well the following relation between CDF and MGF:

$$F_z(Z) = \mathcal{L}^{-1}\left(\frac{\mathcal{M}_z(s)}{s}\right), \quad (6.21)$$

are used, where $\mathcal{L}^{-1}(\cdot)$ denotes the inverse Laplace transformation. Thus, using the table of Laplace transformations, the CDF of the variable a can be obtained as [82]:

$$F_a(x) = 1 - \sum_{j=0}^{k-1}\sum_{i=0}^{M-k}\alpha_j\delta_i\sqrt{\frac{\lambda_{SR}\lambda_{RD}}{\beta_j\varepsilon_i}}xe^{-0.5xL_{j,i}}K_1(x\sqrt{B_{j,i}}) \quad (6.22)$$

with the coefficient $B_{j,i}$ being previously defined with the relation (5.2.30).

The PDF of the variable a can be found differentiating the derived expression for CDF, and implementing the following rule for differentiation of the second order modified Bessel function of the v-th kind [81, Equation (8.486.12)]:

$$\frac{dK_v(z)}{dz} = -K_{v-1}(z) - \frac{v}{z}K_v(z), \quad (6.23)$$

what leads to:

$$f_a(x) = \sum_{j=0}^{k-1}\sum_{i=0}^{M-k}\frac{\alpha_j\delta_i}{2\bar{\gamma}_{SR}\bar{\gamma}_{RD}}xe^{-0.5xL_{j,i}}\left[\frac{L_{j,i}}{\sqrt{B_{j,i}}}K_1(x\sqrt{B_{j,i}}) + 2K_0(x\sqrt{B_{j,i}})\right], \quad (6.24)$$

where the coefficient $L_{j,i}$ is defined in the following manner:

$$L_{j,i} = \frac{\beta_j}{\bar{\gamma}_{SR}} + \frac{\varepsilon_i}{\bar{\gamma}_{RD}}. \tag{6.25}$$

Using the relation (6.8) established between the PDF functions of the variables a and $\gamma_{k,end}$, the PDF of the k-th subcarrier end-to-end SNR becomes:

$$f_{\gamma_k.end}^{BTW}(x) = 2 \sum_{j=0}^{k-1} \sum_{i=0}^{M-k} \frac{\alpha_j \delta_i}{\bar{\gamma}_{SR} \bar{\gamma}_{RD}} x e^{-x L_{j,i}}$$

$$\times \left[\frac{L_{j,i}}{\sqrt{B_{j,ii}}} K_1(2x \sqrt{B_{j,i}}) + 2K_0(2x \sqrt{B_{j,i}}) \right], \tag{6.26}$$

or, given in expanded form:

$$f_{\gamma_k.end}^{BTW}(x) = 2 \sum_{j=0}^{k-1} \sum_{i=0}^{M-k} \frac{\alpha_j \alpha_i}{\bar{\gamma}_{SR} \bar{\gamma}_{RD}} x e^{-x\left(\frac{\beta_j}{\bar{\gamma}_{SR}} + \frac{\varepsilon_i}{\bar{\gamma}_{RD}}\right)} \left[\frac{\bar{\gamma}_{RD}\beta_j + \bar{\gamma}_{SR}\varepsilon_i}{\sqrt{\bar{\gamma}_{SR} \bar{\gamma}_{RD} \varepsilon_i \beta_j}} K_1 \right.$$

$$\times \left. \left(2x \sqrt{\frac{\beta_j \varepsilon_i}{\bar{\gamma}_{SR} \bar{\gamma}_{RD}}} \right) + 2K_0 \left(2x \sqrt{\frac{\beta_j \varepsilon_i}{\bar{\gamma}_{SR} \bar{\gamma}_{RD}}} \right) \right]. \tag{6.27}$$

Figure 6.2 shows the PDFs of SNR obtained for different subcarriers at the receiving end of the analyzed relay system with $M = 16$ subcarriers, with the value of the average SNR per hop being equal to 10 dB. It can be observed that the average received SNR increases with the increase of the subcarrier order, from $k = 1$ to $k = 8$, when the BTW SCP scheme is implemented. On the other side, the average received SNR decreases with the further increase of the subcarrier order, for k being between 9 and 16. It is also important to notice from Figure 6.2, that the PDFs of the received SNR are identical for subcarriers of the orders i and j, if the following condition is satisfied:

$$i + j = M + 1. \tag{6.28}$$

Thus, it is clear that these subcarriers will be characterized with the same BER values, as well as with the same capacity.

6.2.3 PDF of SNR for BTB SCP Scheme

The end-to-end SNR probability density function for the BTB SCP scheme can be derived using the same approach as in the previously described case of

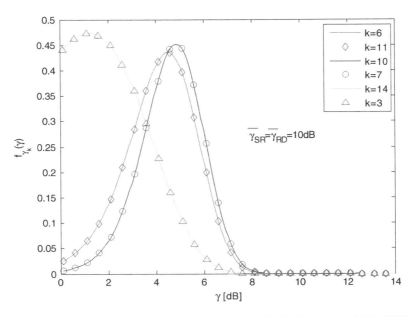

Fig. 6.2 PDF of SNR for the k-th subcarrier at D in the OFDM AF FG relay system with the BTW SCP scheme.

the BTW SCP scheme in OFDM AF VG relay systems. The only difference is in the subcarrier permutation algorithm. Here, subcarriers from the S-RS link are first sorted at the relay station in increasing order with respect to their instantaneous SNRs. Then, they are mapped with the subcarriers of the RS-D link, which are also sorted in increasing order relative to their instantaneous SNRs. In those circumstances, the following expression for the PDF of the k-th subcarrier SNR at the receiving end of the system with BTB SCP is obtained:

$$f_{\gamma_k,end}^{BTB}(x) = 2\sum_{j=0}^{k-1}\sum_{i=0}^{k-1} \frac{\alpha_j \alpha_i}{\bar{\gamma}_{SR}\bar{\gamma}_{RD}} x e^{-x\left(\frac{\beta_j}{\bar{\gamma}_{SR}}+\frac{\beta_i}{\bar{\gamma}_{RD}}\right)} \left[\frac{\bar{\gamma}_{RD}\beta_j + \bar{\gamma}_{SR}\beta_i}{\sqrt{\bar{\gamma}_{SR}\bar{\gamma}_{RD}\beta_i\beta_j}} \right.$$

$$\left. \times K_1\left(2x\sqrt{\frac{\beta_j\beta_i}{\bar{\gamma}_{SR}\bar{\gamma}_{RD}}}\right) + 2K_0\left(2x\sqrt{\frac{\beta_j\beta_i}{\bar{\gamma}_{SR}\bar{\gamma}_{RD}}}\right)\right], \quad (6.29)$$

Introducing the coefficient $I_{j,i}$:

$$I_{j,i} = \frac{\beta_j}{\bar{\gamma}_{SR}} + \frac{\beta_i}{\bar{\gamma}_{RD}}, \quad (6.30)$$

and using the coefficient $A_{j,i}$, previously given with the relation (5.45), the expression (6.29) can be modified into:

$$f^{BTB}_{\gamma_k,end}(x) = 2 \sum_{j=0}^{k-1} \sum_{i=0}^{k-1} \frac{\alpha_j \alpha_i}{\bar{\gamma}_{SR} \bar{\gamma}_{RD}} x e^{-x I_{j,i}}$$

$$\times \left[\frac{I_{j,i}}{\sqrt{A_{j,i}}} K_1 (2x \sqrt{A_{j,i}}) + 2 K_0 (2x \sqrt{A_{j,i}}) \right]. \quad (6.31)$$

Figure 6.3 represents PDFs of SNR for different subcarriers k at the receiving end of the OFDM AF VG relay system with BTB SCP, with the average SNRs on both hops being equal to 10 dB. It is obvious that the mean value of SNR at D becomes higher as the subcarrier order k increases. Comparing the presented graphs with the ones given for the OFDM AF FG relay system shown in Figure 5.3, it can be concluded that the subcarrier mean SNR values at D mutually less differ when VG is implemented. At the same time for all subcarriers, in the case of VG, PDF of SNR graphs are more

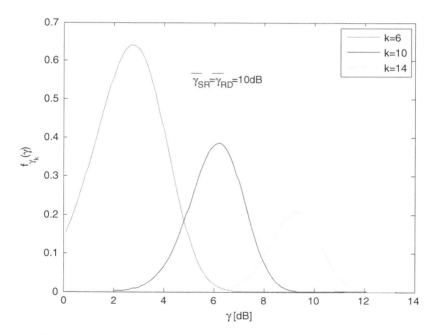

Fig. 6.3 PDF of SNR for the k-th subcarrier at D in the OFDM AF FG relay system with the BTB SCP scheme.

concentrated around their respective mean SNR values, so that the probability of having SNR equal or close to the mean SNR value is higher than for FG systems.

6.2.4   MGF of SNR for BTW SCP Scheme

The MGF of the k-th subcarrier end-to-end SNR is defined with:

$$\mathcal{M}_{\gamma_k,end}(s) = \mathbf{E}(e^{-s\gamma}) = \int\limits_0^\infty f^{BTW}_{\gamma_k,end}(\gamma)e^{-s\gamma}d\gamma. \qquad (6.32)$$

Introducing the previously derived expression for the PDF of the k-th subcarrier received SNR and the BTW SCP scheme (6.26), into the above relation (6.32), the following is obtained:

$$\mathcal{M}_{\gamma_k,end}(s) = \frac{2}{\bar{\gamma}_{SR}\bar{\gamma}_{RD}} \sum_{j=0}^{k-1}\sum_{i=0}^{M-k} \alpha_j\delta_i(\mathcal{I}_1 + \mathcal{I}_2), \qquad (6.33)$$

where:

$$\mathcal{I}_1 = \int_0^\infty \gamma e^{-\gamma(s+L_{j,i})}\frac{L_{j,i}}{\sqrt{B_{j,i}}}K_1(2\gamma\sqrt{B_{j,i}})d\gamma \qquad (6.34)$$

and

$$\mathcal{I}_2 = 2\int\limits_0^\infty \gamma e^{-\gamma(s+L_{j,i})}K_0(2\gamma\sqrt{B_{j,i}})d\gamma. \qquad (6.35)$$

Further on, it is possible to find closed form solutions of the above given integrals, using the integral described in [81]:

$$\int_0^\infty x^{\mu-1}e^{-\alpha x}K_\nu(\beta x)dx = \frac{\sqrt{\pi}(2\beta)^\nu}{(\alpha+\beta)^{\mu+\nu}}\frac{\Gamma(\mu+\nu)\Gamma(\mu-\nu)}{\Gamma(\mu+1/2)}$$
$$\times {}_2F_1\left(\mu+\nu,\nu+\frac{1}{2};\mu+\frac{1}{2};\frac{\alpha-\beta}{\alpha+\beta}\right),$$
$$(\mathrm{Re}\mu) > |\mathrm{Re}(\nu)|, \mathrm{Re}(\alpha+\beta) > 0) \qquad (6.36)$$

Thus, the expression for the MGF of the k-th subcarrier received SNR and the BTW SCP mapping, can be obtained in the following form:

$$\mathcal{M}_{\gamma_{k,end}}(s) = \frac{16}{3} \sum_{j=0}^{k-1} \sum_{i=0}^{M-k} \frac{\alpha_j \delta_i}{\bar{\gamma}_{SR}\bar{\gamma}_{RD}} \frac{1}{(s + L_{j,i} + 2\sqrt{B_{j,i}})^2}$$

$$\times \left[\frac{4L_{j,i}}{s + L_{j,i} + 2\sqrt{B_{j,i}}} {}_2F_1\left(3, \frac{3}{2}; \frac{5}{2}; \frac{s + L_{j,i} - 2\sqrt{B_{j,i}}}{s + L_{j,i} + 2\sqrt{B_{j,i}}}\right) \right.$$

$$\left. + {}_2F_1\left(3, \frac{3}{2}; \frac{5}{2}; \frac{s + L_{j,i} - 2\sqrt{B_{j,i}}}{s + L_{j,i} + 2\sqrt{B_{j,i}}}\right) \right], \tag{6.37}$$

where ${}_2F_1(\cdot, \cdot; \cdot; \cdot)$ denotes the standard Gaussian hypogeometric function:

$$_2F_1(a, b; c; z) = \frac{\Gamma(c)}{\Gamma(a)\Gamma(b)} \sum_{n=0}^{\infty} \frac{\Gamma(a+n)\Gamma(b+n)}{\Gamma(c+n)} \frac{z^n}{n}. \tag{6.38}$$

6.2.5 MGF of SNR for BTB SCP Scheme

Applying the same steps as for the case of the BTW SCP scheme, the expression for the MGF of the k-th subcarrier received SNR when the BTB SCP scheme is implemented can be obtained by introducing the appropriate PDF of SNR given with relation (6.29) into the relation for the MGF of SNR (6.32), and using the integral defined in (6.36). In that manner it becomes:

$$\mathcal{M}_{\gamma_{k,end}}(s) = \frac{16}{3} \sum_{j=0}^{k-1} \sum_{i=0}^{k-1} \frac{\alpha_j \alpha_i}{\bar{\gamma}_{SR}\bar{\gamma}_{RD}} \frac{1}{(s + I_{j,i} + 2\sqrt{A_{j,i}})^2}$$

$$\times \left[\frac{4I_{j,i}}{s + I_{j,i} + 2\sqrt{A_{j,i}}} {}_2F_1\left(3, \frac{3}{2}; \frac{5}{2}; \frac{s + I_{j,i} - 2\sqrt{A_{j,i}}}{s + I_{j,i} + 2\sqrt{A_{j,i}}}\right) \right.$$

$$\left. + {}_2F_1\left(3, \frac{3}{2}; \frac{5}{2}; \frac{s + I_{j,i} - 2\sqrt{A_{j,i}}}{s + I_{j,i} + 2\sqrt{A_{j,i}}}\right) \right]. \tag{6.39}$$

The expressions derived in this paragraph for PDFs and MGFs of the received SNR per subcarrier, represent a basis for performance determination of OFDM AF VG relay systems with BTW or BTB subcarrier mapping at the relay station. Thus, the closed form relations for BER, for both coherent and noncoherent modulations of the information signal, can be defined, as well as the relations describing ergodic capacity of the systems considered.

6.3   BER Performance of OFDM AF VG Relay Systems with SCP

6.3.1   BER Performance of DPSK Modulated OFDM AF VG Relay Systems with SCP

Using the MGF approach, i.e., knowing the MGF of the end-to-end SNR, the BER expression for the k-th received subcarrier, when DPSK modulation is applied, is defined as [82]:

$$P_{b,k} = 0,5\mathcal{M}_{\gamma_{k,end}}(1). \tag{6.40}$$

Introducing the MGF of SNR relation given with (6.37), it becomes possible to obtain the BER expression for the k-th received subcarrier in the OFDM AF VG relay system with the BTW SCP scheme and DPSK modulation. In the same manner, introducing the adequate MGF of SNR relation (6.39), the BER expression for the DPSK modulated OFDM AF VG relay system with the BTB SCP scheme can be derived. Thus, the average BER value can be obtained by averaging BER per subcarrier over the complete subcarrier set:

$$P_b = \frac{1}{M} \sum_{k=1}^{M} P_{b,k}. \tag{6.41}$$

6.3.2   BER Performance of BPSK Modulated OFDM AF VG Relay Systems with SCP

In order to derive BER expression for the BPSK modulated OFDM AF VG relay system with SCP, the PDF approach, as well as the approximation of the complementary error function, are used. Thus, with the same procedure as the one applied in the case of OFDM AF relay systems with fixed gain and SCP, BER for the k-th received subcarrier can be defined with:

$$P_{b,k} = \int_0^\infty P_{b|\gamma_k} f_{\gamma_{k,end}}(\gamma)d\gamma. \tag{6.42}$$

When the BTW SCP scheme is implemented, the above given expression is transformed introducing the adequate relation for the conditional bit error rate given with (5.50) for BPSK modulation and the relation (5.51) for the approximation of the complementary error function. Thus, the expression for bit error rate for the k-th received subcarrier and the BTW subcarrier mapping, becomes:

$$P_{b,k} = \frac{1}{2\bar{\gamma}_{SR}\bar{\gamma}_{RD}} \sum_{j=0}^{k-1} \sum_{i=0}^{M-k} \alpha_j \delta_i (\mathcal{I}_1 + \mathcal{I}_2 + \mathcal{I}_3 + \mathcal{I}_4). \tag{6.43}$$

Integrals $\mathcal{I}_1, \mathcal{I}_2, \mathcal{I}_3, \mathcal{I}_4$ are defined with:

$$\mathcal{I}_1 = \frac{1}{3} \frac{L_{j,i}}{B_{j,i}} \int_0^\infty \gamma e^{-\gamma(1+L_{j,i})} K_1(2\gamma\sqrt{B_{j,i}}) d\gamma, \tag{6.44}$$

$$\mathcal{I}_2 = \frac{L_{j,i}}{\sqrt{B_{j,i}}} \int_0^\infty \gamma e^{-\gamma(4/3+L_{j,i})} K_1(2\gamma\sqrt{B_{j,i}}) d\gamma, \tag{6.45}$$

$$\mathcal{I}_3 = \frac{2}{3} \int_0^\infty \gamma e^{-\gamma(1+L_{j,i})} K_0(2\gamma\sqrt{B_{j,i}}) d\gamma, \tag{6.46}$$

$$\mathcal{I}_4 = 2 \int_0^\infty \gamma e^{-\gamma(4/3+L_{j,i})} K_0(2\gamma\sqrt{B_{j,i}}) d\gamma. \tag{6.47}$$

After some mathematical transformations, each one of the above given integrals can be modified into the standard integral previously described with the relation (6.36). Thus, the solution of the integral $\mathcal{I}_1$ is obtained in the following form:

$$\mathcal{I}_1 = \frac{1}{3} \frac{4\sqrt{\pi} L_{j,i}}{(1 + L_{j,i} + 2\sqrt{B_{j,i}})^3} \frac{\Gamma(3)\Gamma(1)}{\Gamma(5/2)} {}_2F\left(3, \frac{3}{2}; \frac{5}{2}; \frac{1 + L_{j,i} - 2\sqrt{B_{j,i}}}{1 + L_{j,i} + 2\sqrt{B_{j,i}}}\right). \tag{6.48}$$

Further modifications of the above expression can be performed using the known properties of Gamma function:

$$\Gamma(n+1) = n! \Rightarrow \Gamma(3) = 2! = 2,$$

$$\Gamma\left(n + \frac{1}{2}\right) = \frac{1 \cdot 3 \cdot \cdots \cdot (2n-1)}{2^n} \Gamma(1/2) \Rightarrow \Gamma\left(\frac{5}{2}\right) = 3\sqrt{\pi}/4, \tag{6.49}$$

where $\Gamma(1/2) = \sqrt{\pi}$. Finally, the solution of the integral $\mathcal{I}_1$ becomes:

$$\mathcal{I}_1 = \frac{32}{9} \frac{L_{j,i}}{(1 + L_{j,i} + 2\sqrt{B_{j,i}})^3} {}_2F\left(3, \frac{3}{2}; \frac{5}{2}; \frac{1 + L_{j,i} - 2\sqrt{B_{j,i}}}{1 + L_{j,i} + 2\sqrt{B_{j,i}}}\right). \tag{6.50}$$

Applying the same procedure, solutions of the integrals $\mathcal{I}_2, \mathcal{I}_3$ and $\mathcal{I}_4$ can be found:

$$\mathcal{I}_2 = \frac{32}{3} \frac{L_{j,i}}{(4/3 + L_{j,i} + 2\sqrt{B_{j,i}})^3} {}_2F\left(3, \frac{3}{2}; \frac{5}{2}; \frac{4/3 + L_{j,i} - 2\sqrt{B_{j,i}}}{4/3 + L_{j,i} + 2\sqrt{B_{j,i}}}\right),$$

$$(6.51)$$

$$\mathcal{I}_3 = \frac{8}{9} \frac{1}{(1 + L_{j,i} + 2\sqrt{B_{j,i}})^2} {}_2F\left(2, \frac{1}{2}; \frac{5}{2}; \frac{1 + L_{j,i} - 2\sqrt{B_{j,i}}}{1 + L_{j,i} + 2\sqrt{B_{j,i}}}\right), \quad (6.52)$$

$$\mathcal{I}_4 = \frac{8}{3} \frac{1}{(4/3 + L_{j,i} + 2\sqrt{B_{j,i}})^2} {}_2F\left(2, \frac{1}{2}; \frac{5}{2}; \frac{4/3 + L_{j,i} - 2\sqrt{B_{j,i}}}{4/3 + L_{j,i} + 2\sqrt{B_{j,i}}}\right).$$

$$(6.53)$$

Introducing relations (6.50), (6.51), (6.52) and (6.53) into the relation (6.43), the final closed form BER expression for the k-th subcarrier at the receiving end of the OFDM AF VG relay system with BTW SCP , and BPSK modulation, is determined.

When the BTB SCP scheme is implemented, bit error rate for the k-th received subcarrier can be calculated as:

$$P_{b,k} = \frac{1}{2\bar{\gamma}_{SR}\bar{\gamma}_{RD}} \sum_{j=0}^{k-1} \sum_{i=0}^{k-1} \alpha_j \alpha_i (\mathcal{I}_1 + \mathcal{I}_2 + \mathcal{I}_3 + \mathcal{I}_4) \qquad (6.54)$$

Integrals from the above relation (6.54), $\mathcal{I}_1, \mathcal{I}_2, \mathcal{I}_3$ and $\mathcal{I}_4$, can be solved applying the same procedure as the one described for the case of the BTW SCP scheme. Thus, their solutions can be found in the following form:

$$\mathcal{I}_1 = \frac{32}{9} \frac{I_{j,i}}{(1 + I_{j,i} + 2\sqrt{A_{j,i}})^3} {}_2F\left(3, \frac{3}{2}; \frac{5}{2}; \frac{1 + I_{j,i} - 2\sqrt{A_{j,i}}}{1 + I_{j,i} + 2\sqrt{A_{j,i}}}\right), \quad (6.55)$$

$$\mathcal{I}_2 = \frac{32}{3} \frac{I_{j,i}}{(4/3 + I_{j,i} + 2\sqrt{A_{j,i}})^3} {}_2F\left(3, \frac{3}{2}; \frac{5}{2}; \frac{4/3 + I_{j,i} - 2\sqrt{A_{j,i}}}{4/3 + I_{j,i} + 2\sqrt{A_{j,i}}}\right),$$

$$(6.56)$$

$$\mathcal{I}_3 = \frac{8}{9} \frac{1}{(1 + I_{j,i} + 2\sqrt{A_{j,i}})^2} {}_2F\left(2, \frac{1}{2}; \frac{5}{2}; \frac{1 + I_{j,i} - 2\sqrt{A_{j,i}}}{1 + I_{j,i} + 2\sqrt{A_{j,i}}}\right), \quad (6.57)$$

$$\mathcal{I}_4 = \frac{8}{3} \frac{1}{(4/3 + I_{j,i} + 2\sqrt{A_{j,i}})^2} {}_2F\left(2, \frac{1}{2}; \frac{5}{2}; \frac{4/3 + I_{j,i} - 2\sqrt{A_{j,i}}}{4/3 + I_{j,i} + 2\sqrt{A_{j,i}}}\right).$$

(6.58)

By averaging the derived relation for the BER of the k-th received subcarrier (6.43), or (6.54), over the complete set of M subcarriers, the average BER for the OFDM AF VG relay system with BTW SCP, or BTB SCP, is determined.

6.4 Ergodic Capacity of OFDM AF VG Relay Systems with SCP

Ergodic capacity of the k-th received subcarrier, for the analyzed relay system where the data transmission toward the terminal D is performed in two time slots, can be described with:

$$C_k = \frac{1}{2}\mathbf{E}(\log_2(1 + \gamma)) = \frac{1}{2}\int_0^\infty \log_2(1 + \gamma)f_{\gamma_{k,end}}(\gamma)d\gamma,$$

(6.59)

where $f_{\gamma_{k,end}} \in \{f_{\gamma_{k,end}}^{BTW}, f_{\gamma_{k,end}}^{BTB}\}$. Due to the fact that, for the both BTW SCP and BTB SCP schemes, the PDF of the end-to-end SNR is described with the second order modified Bessel functions, it becomes impossible to find closed form analytical solutions of the integral given in the relation (6.59). Therefore, as it is the case with the OFDM AF FG relay system, the approach defining the upper bound values of ergodic capacity can be adopted. In that sense, the Jensen's inequality given in (5.66), and the fact that the log function is a concave function, enables a modification leading toward:

$$C_k \leq \frac{1}{2}\log_2(1 + \mathbf{E}(\gamma)).$$

(6.60)

The expectation $\mathbf{E}(\gamma)$ in the relation (6.60) for the BTW SCP scheme, can be written in the following form:

$$\mathbf{E}(\gamma) = \int_0^\infty \gamma f_{\gamma_{k,end}}(\gamma)d\gamma = \frac{2}{\bar{\gamma}_{SR}\bar{\gamma}_{RD}}\sum_{j=0}^{k-1}\sum_{i=0}^{M-k}\alpha_j\delta_i(\mathcal{I}_1 + \mathcal{I}_2),$$

(6.61)

where integrals $\mathcal{I}_1$, $\mathcal{I}_2$ are defined as:

$$\mathcal{I}_1 = \frac{L_{j,i}}{\sqrt{B_{j,i}}}\int_0^\infty \gamma^2 e^{-\gamma L_{j,i}}K_1(2\gamma\sqrt{B_{j,i}})d\gamma,$$

(6.62)

$$\mathcal{I}_2 = 2\int_0^\infty \gamma^2 e^{-0.5\gamma L_{j,i}}K_0(2\gamma\sqrt{B_{j,i}})d\gamma$$

(6.63)

Both integrals given with relations (6.62) and (6.63) can be modified into the form of the standard integral, described in the relation (6.36). On that basis, the expression for $\mathbf{E}(\gamma)$, in the case of the BTW SCP scheme applied at RS, can be determined as :

$$
\mathbf{E}(\gamma) = \frac{128}{15\bar{\gamma}_{SR}\bar{\gamma}_{RD}} \sum_{j=0}^{k-1} \sum_{i=0}^{M-k} \frac{\alpha_j \delta_i}{(L_{j,i} + 2\sqrt{B_{j,i}})^3}
$$

$$
\times \left[\frac{3L_{j,i}}{L_{j,i} + 2\sqrt{B_{j,i}}} {}_2F_1\left(4, \frac{3}{2}; \frac{7}{2}; \frac{L_{j,i} - 2\sqrt{B_{j,i}}}{L_{j,i} + 2\sqrt{B_{j,i}}}\right) \right.
$$

$$
\left. + {}_2F_1\left(3, \frac{1}{2}; \frac{7}{2}; \frac{L_{j,i} - 2\sqrt{B_{j,i}}}{L_{j,i} + 2\sqrt{B_{j,i}}}\right) \right]. \tag{6.64}
$$

For the BTB SCP scheme, applying the identical procedure as the one described for the BTW SCP, $\mathbf{E}(\gamma)$ is obtained as:

$$
\mathbf{E}(\gamma) = \frac{128}{15\bar{\gamma}_{SR}\bar{\gamma}_{RD}} \sum_{j=0}^{k-1} \sum_{i=0}^{k-1} \frac{\alpha_j \alpha_i}{\left(I_{j,i+2\sqrt{A_{j,i}}}\right)^3}
$$

$$
\times \left[\frac{3I_{j,i}}{I_{j,i} + 2\sqrt{A_{j,i}}} {}_2F_1\left(4, \frac{3}{2}; \frac{7}{2}; \frac{I_{j,i} - 2\sqrt{A_{j,i}}}{I_{j,i} + 2\sqrt{A_{j,i}}}\right) + \right.
$$

$$
\left. + {}_2F_1\left(3, \frac{1}{2}; \frac{7}{2}; \frac{I_{j,i} - 2\sqrt{A_{j,i}}}{I_{j,i} + 2\sqrt{A_{j,i}}}\right) \right] \tag{6.65}
$$

Including the derived expression for $\mathbf{E}(\gamma)$ (6.64), or (6.65), into the relation (6.60), the upper bound of the achievable ergodic capacity for the OFDM AF VG relay system with BTW SCP, or BTB SCP scheme, can be determined.

6.5   Performance Analysis of OFDM AF VG Relay Systems With SCP

This section gives detailed analyses of BER and ergodic capacity performances, for OFDM AF VG relay systems with SCP. A part from presenting the analytically obtained results, their verification through comparisons with the simulation results, is performed. The system parameters are assumed to be identical with the ones used for the performance analyses of OFDM

AF FG relay systems with SCP. This means that the system is ideally time and frequency synchronized, with $M = 16$ subcarriers (subcarrier groups) and SCP (BTW or BTB) implemented at the relay station. The average symbol energy emitted by the relay station is taken to be equal to the average symbol energy emitted by the terminal S, i.e., $\epsilon_R = \epsilon_S$, and noise variances at the relay station and the terminal D are also identical, $\mathcal{N}_{01} = \mathcal{N}_{02} = \mathcal{N}_0$. If it is not otherwise stated, S-RS and RS–D distances are assumed to be equal, implying that $\bar{\gamma}_{SR} \cong \bar{\gamma}_{RD}$.

Simulations are performed with the same parameters used and described previously for OFDM AF FG relay systems.

6.5.1 BER Performance Analysis of DPSK Modulated OFDM AF VG Relay Systems with SCP

Figure 6.4 presents BER performance of the OFDM AF VG relay system with SCP and DPSK modulated input data stream. Some of the presented graphs are obtained analytically using the previously explained MGF approach, while the others illustrate the results of adequate simulation procedures. In order to

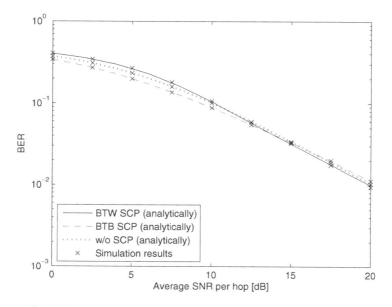

Fig. 6.4 BER performance of DPSK modulated OFDM AF VG relay system.

identify benefits introduced through SCP implementation at the relay station, Figure 6.4 includes BER graphs given for the OFDM AF VG relay system without (w/o) SCP [72].

The BER values given in Figure 6.4 clearly show that all analytically obtained results are completely proven by simulations. In other words, it can be concluded that the algorithms introduced for the end-to-end SNR statistics (PDF and MGF) determination can be considered as valid. Further on, it is obvious that the SCP implementation does not offer significant BER performance enhancements for this type of relay systems, when compared with the improvements achieved in the OFDM AF FG relay systems. Additionally, for the relay system with variable gain, the best BER values are attained with the BTB SCP scheme when average SNR per hop is bellow 13 dB, while above this SNR value the advantage is on the side of the BTW SCP. Comparing the illustrated results for the two SCP schemes, it can be noted that for the BER value of 10^{-2}, the system with BTW SCP scheme has a SNR gain of around 0.5 dB, while for BER of 10^{-1} the system with BTB SCP has a SNR gain of 1 dB.

6.5.2   BER Performance Analysis of BPSK Modulated OFDM AF VG Relay Systems with SCP

Figure 6.5 gives BER values, obtained analytically and by simulations, for the BPSK modulated OFDM AF VG relay system with subcarrier permutation implemented at the relay station. The graph showing the simulated BER values for the OFDM AF VG relay system with no subcarrier permutation is also included. Generally, it can be observed that the analytical results and the simulation results are very close. Despite a certain difference among them, the BER values obtained through the previously described approximations can be considered as valid for performance analyses. Precisely, this difference between the simulated and analytically obtained BER values is always less than 0.5 dB, for each BER value considered.

As it is the case with the situation described for the DPSK modulated system, it can be observed that the implementation of a SCP scheme does not enable significant BER performance improvement in BPSK modulated OFDM AF VG relay system. Additionally, for SNR values per hop bellow 12 dB, the best BER results are achieved with the BTB SCP scheme, while for SNRs above 12 dB it is the BTW SCP scheme which enables better BER

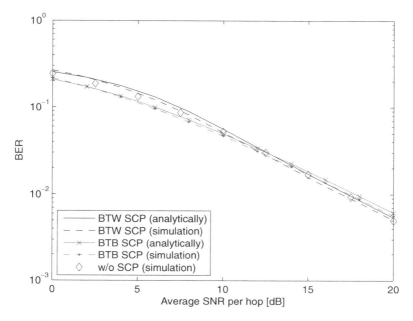

Fig. 6.5 BER performance of BPSK modulated OFDM AF VG relay system.

performance. The illustrated graphs show that for the BER value of 10^{-2}, the BTW SCP scheme has a SNR gain of 0.5 dB in comparison with the BTB SCP scheme, and around 0.25 dB when compared with the system where no SCP is implemented.

Generally, when analyzing the presented BER performance of DPSK and the BPSK modulated OFDM AF VG relay systems with subcarrier permutation, it can be concluded that only the implementation of the BTB SCP scheme can be justified from the point of the system complexity. When compared with the less complex system with no subcarrier permutation, the system with BTB SCP has better BER performance in the range of low average SNR values, i.e., when the channel propagation conditions can be considered as unfavorable for the deployment of the system with no SCP.

6.5.3 Ergodic Capacity Analysis of OFDM AF VG Relay Systems with SCP

Analytically determined values for the upper bound of the achievable ergodic capacity per subcarrier at the receiving end of the OFDM AF VG relay system

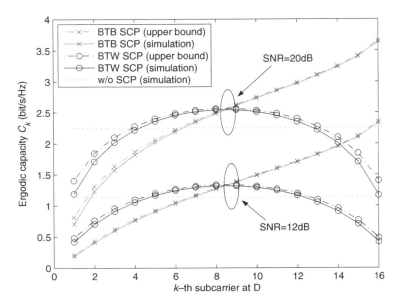

Fig. 6.6   Ergodic capacity per subcarrier for OFDM AF VG relay system.

with BTB SCP or BTW SCP, for the average SNR values on both hops of 12 dB and 20 dB, are presented in Figure 6.6. Results for the ergodic capacity obtained through simulations are also given, as well as ergodic capacity values for the OFDM AF VG system with no subcarrier permutation.

When performing simulations, the instantaneous value of the ergodic capacity for the k-th subcarrier in the n-th channel realization is defined with:

$$C_k^{(n)} = \frac{1}{2} \log_2 \left(1 + \frac{(G_k^{(n)})^2 |H_{1,k}^{(n)} H_{2,k}^{(n)} X_k^{(n)}|^2}{(G_k^{(n)})^2 |H_{2,k}^{(n)}|^2 \mathcal{N}_{0_1} + \mathcal{N}_{0_2}} \right). \tag{6.66}$$

Applying the same procedure as in the case of OFDM AF FG relay systems, independent transfer functions having the Rayleigh fading distributions are generated for each subcarrier on the S-RS link ($H_{1,k}^{(n)}$) and on the RS-D link ($H_{2,k}^{(n)}$), for each n-th channel realization, and $1 \leq n \leq n_{tot}$. The ergodic capacity for the k-th subcarrier is then determined averaging the ergodic capacities obtained for particular channel realization:

$$C_k = \frac{1}{n_{tot}} \sum_{n=1}^{n_{tot}} C_k^{(n)}. \tag{6.67}$$

Table 6.1. Average ergodic capacity per subcarrier at the receiving end of the OFDM AF VG relay system.

	SNR = 0 dB		SNR = 12 dB		SNR = 20 dB	
	Simul.	Analyt.	Simul.	Analyt.	Simul.	Analyt.
BTW SCP [b/s/Hz]	0,144	0,148	1,014	1,052	2,124	2,202
BTB SCP [b/s/Hz]	0,250	0,2502	1,292	1,310	2,429	2,464
No SCP [b/s/Hz]	0,192		1,137		2,256	

Figure 6.6 proves that the analytical approach presented in Section 6.4 gives the upper bound of the achievable ergodic capacity per subcarrier which is very close to the values obtained through simulations. More clear insight into the quality of the approximations introduced in the analytical procedures, and into the relations among ergodic capacities for the particular SCP schemes, can be obtained if the average achievable ergodic capacity per subcarrier is calculated for: the system with BTW SCP, the system with BTB SCP and the system with no subcarrier permutation (Table 6.1).

Thus, in comparison with the system with no SCP, the BTB SCP scheme enables an increase of the ergodic capacity of 30%, 13,6% and 7,6% for the average SNR per hop of 0 dB, 12 dB and 20 dB, respectively. It is worth noticing, that the capacity enhancements achieved with BTB SCP scheme are significant, especially in the range of lower average SNRs per hop.

If there would be a possibility that the OFDM AF VG relay system makes a switch from the BTB SCP scheme to the BTW SCP scheme, when average SNR values on both hops become equal to 12 dB, corresponding with the SNR value above which BTW SCP achieves the best BER performance for the BPSK modulation, then the capacity loss would be 21,5%. Also, the capacity loss with BTW SCP relative to the system with no SCP, would be 10,8%, for the same average SNR on both hops. With the increase of average SNR, the difference between BTW SCP and the other two analyzed schemes becomes smaller. Thus, for example for the average SNR value on both hops of 20 dB, BTW SCP achieves 12,5% lower capacity than the BTB SCP scheme and 5,8% lower than the system with no SCP.

Generally, it can be concluded that the performance analysis of the OFDM AF VG relay system with SCP, from the point of the achievable capacity, additionally approves the conclusions obtained through their BER performance analysis. Thus, it can be stated that the increased system complexity due to the implementation of the BTB SCP scheme can be considered as justified, since this SCP scheme enables significant capacity improvements for all values of SNR on both hops, and especially in the range of low SNRs. It is the same subcarrier permutation method for which the analyses presented in 6.5.1 and 6.5.2 have shown the ability to enhance BER performance in the range of low and medium SNRs per hop. In the range of higher SNRs, it is the BTW SCP scheme that achieves the best BER results, with the level of improvement that can be almost neglected in comparison with the system with no SCP, eliminating thus the need for its implementation.

6.6 Performance Comparison of OFDM AF Relay Systems with SCP

Identification of an optimal configuration for the relay station in OFDM AF relay systems with SCP might be interesting from the point of a potential solution for the Type II relay stations and their implementation in the next generation WWAN networks. In that context, a comprehensive comparison of OFDM AF relay systems with fixed gain with those with variable gain offers significant set of information, not only when their performances are taken into account but their complexity as well, i.e., the fact that systems with variable gain include a higher number of amplifying blocks (Figures 5.1 and 6.1). Since it has already been shown that the BTB SCP scheme is the superior subcarrier permutation scheme when the achievable ergodic capacity is concerned, the following comparisons will only deal with OFDM AF systems implementing BTB SCP.

In order to perform valid comparisons, an ideally synchronized OFDM AF relay system with $M = 16$ subcarriers is assumed for both FG and VG, with noise variances at RS and D terminal being equal, $\mathcal{N}_{01} = \mathcal{N}_{02} = \mathcal{N}_0$. For the system with fixed gain, the gain factor G is taken to be:

$$G^2 = \frac{e_R}{e_S E |H_{1,k}|^2} e^{1/\bar{\gamma}_{SR}} E_1 \left(\frac{1}{\bar{\gamma}_{SR}} \right), \tag{6.68}$$

ensuring that the average energy per symbol transmitted by RS is equal with the average energy per symbol transmitted by S, i.e., $e_R = e_S$, since it is calculated from:

$$G^2 = \mathbf{E} \left[\frac{e_R}{e_S |H_{1,k}|^2 + \mathcal{N}_0} \right]. \tag{6.69}$$

For the system with variable gain, the gain introduced at the relay station on the k-th subcarrier is given with:

$$G_k = \frac{1}{H_{1,k}}, \tag{6.70}$$

meaning that the average gain of the FG system differs from the gain in the VG system only in the fact that it takes into account the noise power from the first hop. Thus, when the noise power can be neglected in comparison with the signal power, it becomes possible to consider the average gains in both FG and VG systems to be equal.

6.6.1 Comparison of BER Performances

BER performance of DPSK modulated OFDM AF FG and OFDM AF VG relay systems with SCP, when the average SNR on the S-RS link is identical with the average SNR on the RS-D link, are illustrated in Figure 6.7. Their comparison clearly shows that the BTW SCP scheme applied in the AF system

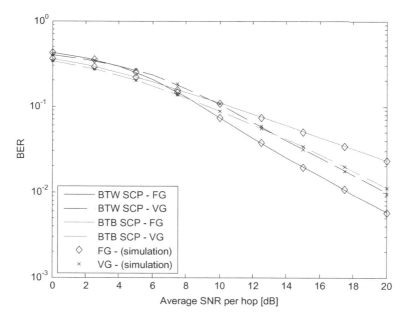

Fig. 6.7 BER performance of DPSK modulated OFDM AF FG and OFDM AF VG relay systems with SCP.

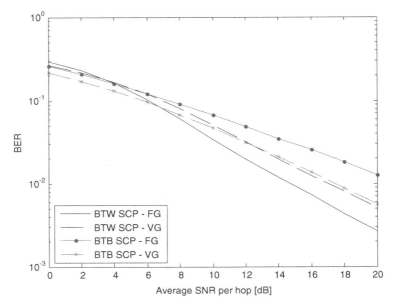

Fig. 6.8 BER performance of BPSK modulated OFDM AF FG and OFDM AF VG relay systems with SCP.

with FG achieves the best BER performance for average SNR per hop values above 8 dB. For example, for BER $= 10^{-2}$ the SNR gain of the FG relay system with BTW SCP scheme is more than 2 dB in comparison with the same scheme applied in the system with variable gain. When average SNRs per hop bellow 8 dB are considered, the VG relay system with BTB SCP has the worst BER performance, offering at the same time a minor performance improvement in comparison with the FG AF system with BTB SCP. The same conclusions stands for BPSK modulated relay systems (Figure 6.8). The only difference is in the fact that the average per hop SNR value above which the FG relay system with BTW SCP can be considered as better, becomes smaller (around 6,5 dB).

It is expected that the next generation WWAN networks will incorporate infrastructure relay stations, with the main task of relaying information between S and D, thus enabling quality of service improvements. In such scenario, a realistic assumption is that the time-variation of the channel between the base station and RS are very small. This implies the instantaneous SNR values on the S-RS link to be close to the average SNR on this hop. Under this

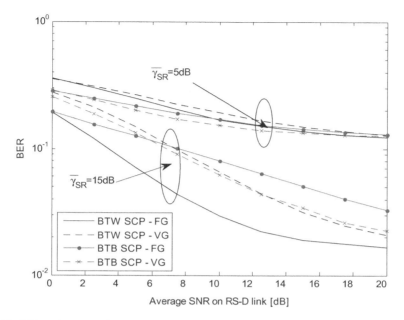

Fig. 6.9 BER performance of DPSK modulated OFDM AF FG and OFDM AF VG relay systems with SCP ($\bar{\gamma}_{SR} = 5$ dB and $\bar{\gamma}_{SR} = 15$ dB).

condition, BER performances are shown in Figure 6.9 (DPSK modulation) and Figure 6.10 (BPSK modulation), for $\bar{\gamma}_{SR} = 5$ dB and $\bar{\gamma}_{SR} = 15$ dB.

In the scenario with the average SNR on the S-RS link being $\bar{\gamma}_{SR} = 15$ dB, the OFDM AF FG relay system with BTW SCP has the best BER performance for all values of SNR on the RS-D link. However, for small average SNR values on the S-RS link, it is the OFDM AF VG relay system with BTB SCP scheme which outperforms the other relay techniques, as presented in Figures 6.9 and 6.10. From the given graphs it can be seen that in the scenario with $\bar{\gamma}_{SR} = 5$ dB, the OFDM AF VG implementing BTB SCP has the best BER performance, for the average SNRs on the RS-D link up to almost 16 dB for both analyzed modulation schemes. The analysis shows that whenever $\bar{\gamma}_{SR}$ is below 8 dB, there appears a region of average SNR values on the RS-D link, where the OFDM AF VG relay system with BTB SCP achieves the lowest BER values.

In the case when criteria for the optimal relay station configuration include a delay introduced at the relay station, then the choice is between FG system and VG system, since they both introduce the same delay which is shorter

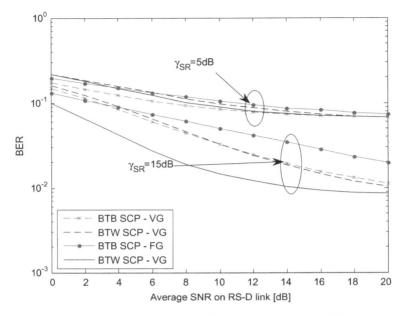

Fig. 6.10 BER performance of BPSK modulated OFDM AF FG and OFDM AF VG relay systems with SCP ($\bar{\gamma}_{SR} = 5\,\text{dB}$ and $\bar{\gamma}_{SR} = 15\,\text{dB}$).

than the one of DF systems. The presented BER performance comparisons of OFDM AF FG and OFDM AF VG relay systems can serve as one of the criteria in indentifying optimal configuration of the Type II relay station for implementation in WWANs. The presented analyses clearly show that it is not possible to define a solution that will be optimal for all SNR values on the hops. Actually, what is possible is to identify a solution that can be considered as an optimal under certain conditions. Namely, for a lower average SNRs on the S-RS link (for example: up to 8 dB for DPSK modulation, up to 6 dB for BPSK modulation), there exists an region of average SNR values on the RS-D link where the OFDM AF VG relay system with BTB SCP can be considered as optimal. Practical examples of such a scenario can include radio coverage inside big buildings, where a smaller relay station can be implemented in the building, or radio coverage of an underground transport station, or emergency radio coverage of limited areas. For all the other, higher average SNRs on the S-RS link, the OFDM AF FG relay system with BTW SCP achieves the best BER performance, irrespective of the SNR value on the RS-D link.

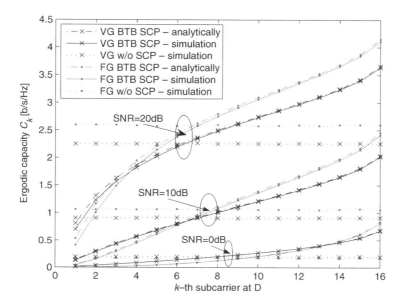

Fig. 6.11 Ergodic capacity per subcarrier for OFDM AF FG and OFDM AF VG relay system with and without SCP.

6.6.2 Comparison of Ergodic Capacities

Figure 6.11 summarizes graphs obtained in previous paragraphs for ergodic capacities per subcarrier at the terminal D, for OFDM AF FG and OFDM AF VG relay systems with BTB SCP, and different average SNR values per hop. Only the results for the BTB SCP scheme are presented, as it has been already shown that, in comparison with the BTW SCP scheme, it achieves the highest capacity values [21–23]. For the sake of comparison, Figure 6.11 includes the ergodic capacity graphs for the relay systems with no subcarrier permutation at RS.

In order to gain a better insight into mutual relations between the shown performances, Table 6.2 gives the values of the average ergodic capacity per subcarrier at the destination terminal. As it can be noticed, analytically obtained upper bounds of the achievable capacity per subcarrier maintain the same ratio among the performance of relay systems with FG and VG as it is the case in the simulation model. Thus, those analytical results can be used in further comparison of the systems considered, when their capacity is the performance of interest. The results summarized in Table 6.2 show that, only in the region

Table 6.2. Average ergodic capacity per subcarrier at D station for OFDM AF FG and OFDM AF VG relay system.

	SNR = 0 dB		SNR = 10 dB		SNR = 20 dB	
	Simul.	Analyt.	Simul.	Analyt.	Simul.	Analyt.
FG- BTB SCP [b/s/Hz]	0,222	0,228	1,144	1,176	2,641	2,702
VG -BTB SCP [b/s/Hz]	0,251	0,252	1,051	1,065	2,429	2,464
FG - no SCP [b/s/Hz]	0,168		1,06		2,595	
VG - no SCP [b/s/Hz]	0,192		0,909		2,256	

Table 6.3. Average ergodic capacity per subcarrier at D station for OFDM AFFG and OFDM AF VG relay system ($\bar{\gamma}_{RD} = 0$ dB).

$\bar{\gamma}_{SR}$ [dB]	1	2	3	4
FG–BTB SCP [b/s/Hz]	0,255	0,289	0,328	0,366
VG–BTB SCP [b/s/Hz]	0,271	0,293	0,312	0,328

of small average SNRs per hop, the relay system with VG achieves higher capacity than the system with FG.

Further analysis of the average ergodic capacities per subcarrier for small average SNRs on both hops shows that the OFDM relay system with FG achieves higher capacity than the system with VG, for all the scenarios with $\bar{\gamma}_{SR} \geq 2,5$ dB (Table 6.3) and $\bar{\gamma}_{RD} \geq 0$ dB.

Since for the real relay system with infrastructure based relay stations, the good condition on the base station-relay station link is of the crucial importance for the successful communication process, the position of the relay station should be carefully chosen. Thus, for the downlink communication process, it is reasonable to assume that the average SNR on the S-RS link is greater than 2,5 dB, and the average SNR on the RS-D link is greater than 0 dB, then it is clear that the OFDM AF FG relay system with BTB SCP can be considered as optimal solution for the Type II relay stations, when the system capacity is the performance of interest.

7

Performance of OFDM DF Relay Systems with Subcarrier Permutation

At the beginning, the idea of subcarrier permutation at the relay station of OFDM based relay systems, enabling an increase of the system achievable capacity as well as BER reductions, was related with systems incorporating amplify and forward signal processing at relay stations. However, as non-transparent relay stations (Type I) in the next generation WWAN networks will be based on DF relaying, it becomes important to determine effects of SCP implementation in OFDM DF relay systems as well.

Following the already known fact that the incorporation of the BTB SCP scheme leads towards significant capacity enhancement [34–36], as well as BER performance improvement [33], of the OFDM DF relay systems, this Chapter gives comprehensive analyses of BER performance and the achievable capacity of OFDM DF relay systems with SCP.

7.1 System Description

The communication scenario with three terminals with no direct communication link between the source S and the destination D is considered. As in the cases of OFDM AF FG and OFDM AF VG relay systems, i.i.d. Rayleigh fading among subcarriers on both hops, and half-duplex operating R station, are assumed.

Block scheme of the R station (RS) performing DF relaying and subcarrier permutation is shown in Figure 7.1. The FFT block for OFDM demodulation is

OFDM Based Relay Systems for Future Wireless Communications, 135–153.

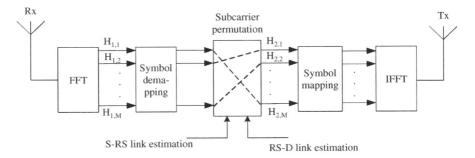

Fig. 7.1 Block scheme of DF relay station with subcarrier permutation.

followed by a block performing symbol demapping. Then, using the estimated subcarrier transfer functions, subcarriers from the first hop are mapped to subcarriers on the second hop. In the next step, symbol mapping is applied and then the OFDM modulation.

The signal received at the relay station, on the i-th subcarrier, after OFDM demodulation can be presented with:

$$Y_{R,i} = X_i H_{1,i} + N_{1,i}, \quad 1 \leq i \leq M. \tag{7.1}$$

The signal on the k-th subcarrier at the D terminal, has the following form in the frequency domain:

$$Y_{D,k} = H_{2,k} \hat{X}_{R,\upsilon(i)} + N_{2,k}, \tag{7.2}$$

where $\hat{X}_{R,\upsilon(i)}$ denotes the estimated symbol received at RS on the i-th subcarrier. It is taken that the function $\upsilon(i)$ maps the i-th subcarrier from the first hop to the k-th subcarrier on the second hop. If subcarriers from the S-RS link are increasingly ordered at RS in accordance to their instantaneous SNRs, then the PDF of SNR for the k-th subcarrier on the S-RS link with Rayleigh fading statistics may be expressed as:

$$f_{k,SR}^w(x) = \sum_{i=0}^{k-1} \frac{1}{\bar{\gamma}_{SR}} \alpha_i e^{-\beta_i x / \bar{\gamma}_{SR}} \tag{7.3}$$

with

$$\alpha_i = (-1)^i M \binom{M-1}{k-1} \binom{k-1}{i} \quad \text{and} \quad \beta_i = i + M - k + 1. \tag{7.4}$$

If subcarriers from the RS-D link are also increasingly ordered at RS in accordance to their instantaneous SNRs, then PDF of SNR for the k-th subcarrier on the RS-D link have the same form as in (7.3), but the appropriate average SNR on the RS-D link, with $\bar{\gamma}_{RD}$ being used instead of $\bar{\gamma}_{SR}$. If the subcarriers on the RS-D link are decreasingly ordered with respect to their instantaneous SNRs, then the PDF of SNR for the k-th subcarrier on the RS-D link can be expressed as:

$$f_{k,RD}^s(x) = \sum_{i=0}^{M-k} \frac{1}{\bar{\gamma}_{RD}} \delta_i e^{-\varepsilon_i x / \bar{\gamma}_{RD}} \tag{7.5}$$

where

$$\delta_i = (-1)^i M \binom{M-1}{k-1} \binom{M-k}{i} \quad \text{and} \quad \varepsilon_i = i + k. \tag{7.6}$$

With the known statistics of SNR for both links involved in the communication process, BER or ergodic capacity performance analyses may be conducted for OFDM DF relay systems implementing BTB SCP or BTW SCP.

7.2 BER Performances of DPSK Modulated OFDM DF Relay Systems with SCP

Following the bit error rate relation (2.37) given for DPSK or BPSK modulated single-carrier relay systems, BER at the receiving end of OFDM DF system for the k-th subcarrier can be found as:

$$P_{b,k} = P_{b1,k} + P_{b2,k} - 2P_{b1,k} P_{b2,k}, \tag{7.7}$$

where $P_{b1,k}$ and $P_{b2,k}$ denote bit error rates for the k-th subcarriers of the first and the second hop, respectively. When DPSK modulation is applied, using the known MGF of SNR, BER for the k−th subcarrier of each hop is determined in the following manner:

$$P_{bi,k} = 0,5\mathcal{M}_{\gamma_k,i}(1), \quad i = 1, 2. \tag{7.8}$$

Assuming that subcarriers on the S-RS link are increasingly ordered in respect to their instantaneous SNRs, and for the scenario with i.i.d. Rayleigh narrowband fading statistics among subcarriers, the PDF of SNR for the k-th subcarrier is given with the relation (7.3). Then, the MGF of SNR for the k-th subcarrier can be defined with:

$$\mathcal{M}_{\gamma_k,1}(s) = \int_0^\infty f_{k,SR}^w(x)e^{-sx}dx = \sum_{i=0}^{k-1} \frac{\alpha_i}{s\bar{\gamma}_{SR} + \beta_i}. \tag{7.9}$$

In order to implement BTW SCP, the subcarriers from the RS-D link have to be decreasingly ordered in respect to their instantaneous SNRs. It means that the PDF of SNR for the k-th subcarrier is defined with the relation (7.5), and the MGF of SNR for the considered subcarrier can be determined as:

$$\mathcal{M}_{\gamma_k,2}(s) = \int_0^\infty f_{k,RD}^S(x)e^{-sx}dx = \sum_{i=0}^{M-k} \frac{\delta_i}{s\bar{\gamma}_{RD} + \varepsilon_i}. \tag{7.10}$$

Introducing relations (7.9) and (7.10) into the relation (7.8), BER expressions for the DPSK modulated k-th subcarrier of the S-RS link and RS-D link, respectively, can be obtained. Then, their combination with the relation (7.7) enables BER determination for the k-th subcarrier at the receiving end of the DPSK modulated OFDM DF relay system with BTW SCP:

$$P_{b,k} = 0,5\left[\sum_{i=0}^{k-1} \frac{\alpha_i}{\bar{\gamma}_{SR} + \beta_i} + \sum_{j=0}^{M-k} \frac{\delta_j}{\bar{\gamma}_{RD} + \varepsilon_j} - \sum_{i=0}^{k-1} \frac{\alpha_i}{\bar{\gamma}_{SR} + \beta_i} \sum_{j=0}^{M-k} \frac{\delta_j}{\bar{\gamma}_{RD} + \varepsilon_j}\right] \tag{7.11}$$

BER of the considered system is obtained by averaging such obtained k-th subcarrier BER values over the set of all M subcarriers, $P_b = \sum_{k=1}^M P_{b,k}/M$.

Using the same approach, BER for OFDM DF system with BTB SCP can be found. In the case of BTB SCP, the MGF of SNR for the k-th subcarrier of the RS-D link is also given with the relation (7.9) (involving the average SNR on the RS-D link instead of $\bar{\gamma}_{SR}$). BER for the k-th subcarrier at the terminal D of the DPSK modulated OFDM DF relay system implementing BTB SCP, is obtained as:

$$P_{b,k} = 0,5\left[\sum_{i=0}^{k-1} \frac{\alpha_i}{\bar{\gamma}_{SR} + \beta_i} + \sum_{j=0}^{M-k} \frac{\alpha_j}{\bar{\gamma}_{RD} + \beta_j} - \sum_{i=0}^{k-1} \frac{\alpha_i}{\bar{\gamma}_{SR} + \beta_i} \sum_{j=0}^{M-k} \frac{\alpha_j}{\bar{\gamma}_{RD} + \beta_j}\right] \tag{7.12}$$

7.3 BER Performances of BPSK Modulated OFDM DF Relay Systems with SCP

When BPSK modulated OFDM DF relay system is considered, the BER relation (7.7) is also valid. Further on, BER for the $k-$th subcarrier of the S-RS link and the RS-D link, can be found using the PDF approach:

$$P_{bi,k} = \int_0^\infty P_{bi|\gamma_k} f_{\gamma_k,i}(x) dx, \quad i = 1, 2. \tag{7.13}$$

For OFDM DF relay system implementing BTW SCP scheme, with the subcarriers on the S-RS link being increasingly ordered in respect to their instantaneous SNRs, $P_{b1,k}$ is obtained in the form:

$$P_{b1,k} = \frac{1}{2\bar{\gamma}_{SR}} \sum_{i=1}^{k-1} \alpha_i \int_0^\infty erfc(\sqrt{x}) e^{-\beta_i x / \bar{\gamma}_{SR}} dx. \tag{7.14}$$

The above given integral can be solved using the following standard integral [81]:

$$\int_0^\infty erfc(\sqrt{a}x) e^{bx} dx = \frac{1}{b} \left(\frac{\sqrt{a}}{\sqrt{a-b}} - 1 \right),$$

$$\left[\text{Re}\{a\} > 0; \text{Re}\{b\} < \text{Re}\{a\} \right]. \tag{7.15}$$

On that basis, $P_{b1,k}$ can be defined as:

$$P_{b1,k} = \frac{1}{2} \sum_{i=0}^{k-1} \frac{\alpha_i}{\beta_i} \left(1 - \frac{2}{\sqrt{1 + \beta_i / \bar{\gamma}_{SR}}} \right). \tag{7.16}$$

Using the same procedure, BER for the k-th subcarrier of the RS-D link can be found, taking into account that the subcarriers on this link are decreasingly ordered in respect to their instantaneous SNRs. Thus, it can be obtained:

$$P_{b2,k} = \frac{1}{2} \sum_{i=0}^{M-k} \frac{\delta_i}{\varepsilon_i} \left(1 - \frac{2}{\sqrt{1 + \varepsilon_i / \bar{\gamma}_{RD}}} \right). \tag{7.17}$$

Introducing relations (7.16) and (7.17) into the relation (7.7), the final BER expression for the k-th subcarrier at the terminal D of the BPSK modulated

OFDM DF relay system with BTW SCP scheme is derived:

$$
P_{b,k} = \frac{1}{2} \left[\sum_{i=0}^{k-1} \frac{\alpha_i}{\beta_i} \left(1 - \frac{2\sqrt{\bar{\gamma}_{SR}}}{\sqrt{\bar{\gamma}_{SR} + \beta_i}} \right) + \sum_{j=0}^{M-k} \frac{\delta_j}{\varepsilon_j} \left(1 - \frac{2\sqrt{\bar{\gamma}_{RD}}}{\sqrt{\bar{\gamma}_{RD} + \varepsilon_j}} \right) \right.
$$
$$
\left. - \sum_{i=0}^{k-1} \frac{\alpha_i}{\beta_i} \left(1 - \frac{2\sqrt{\bar{\gamma}_{SR}}}{\sqrt{\bar{\gamma}_{SR} + \beta_i}} \right) \sum_{j=0}^{M-k} \frac{\delta_j}{\varepsilon_j} \left(1 - \frac{2\sqrt{\bar{\gamma}_{RD}}}{\sqrt{\bar{\gamma}_{RD} + \varepsilon_j}} \right) \right] \quad (7.18)
$$

In the case of the BTB SCP scheme applied, BER for the k-th subcarrier at D in the considered OFDM DF relay system can be calculated as:

$$
P_{b,k} = \frac{1}{2} \left[\sum_{i=0}^{k-1} \frac{\alpha_i}{\beta_i} \left(1 - \frac{2\sqrt{\bar{\gamma}_{SR}}}{\sqrt{\bar{\gamma}_{SR} + \beta_i}} \right) + \sum_{j=0}^{M-k} \frac{\alpha_j}{\beta_j} \left(1 - \frac{2\sqrt{\bar{\gamma}_{RD}}}{\sqrt{\bar{\gamma}_{RD} + \beta_j}} \right) \right.
$$
$$
\left. - \sum_{i=0}^{k-1} \frac{\alpha_i}{\beta_i} \left(1 - \frac{2\sqrt{\bar{\gamma}_{SR}}}{\sqrt{\bar{\gamma}_{SR} + \beta_i}} \right) \sum_{j=0}^{M-k} \frac{\alpha_j}{\beta_j} \left(1 - \frac{2\sqrt{\bar{\gamma}_{RD}}}{\sqrt{\bar{\gamma}_{RD} + \beta_j}} \right) \right] \quad (7.19)
$$

Finally, BER of the BPSK modulated OFDM DF relay system with SCP system is obtained by averaging such obtained k-th subcarrier BER values over the set of all M subcarriers, $P_b = \sum_{k=1}^{M} P_{b,k}/M$.

7.4 Ergodic Capacity of OFDM DF Relay Systems with SCP

When OFDM DF relay system with SCP is considered, with the i-th subcarrier of the S-RS link being mapped to the k-th subcarrier of the RS-D link, the capacity of the k-th received subcarrier depends only on the instantaneous SNR of the "weaker" of the two subcarriers. There the term "weaker" implies the subcarrier with the lower instantaneous SNR. In analyzing ergodic capacity of the OFDM DF relay system two different scenarios can be assumed: (1) a scenario with the equal average SNRs on both hops and (2) a scenario with unequal average SNRs on hops. For the first scenario a closed-form analytical solution for the ergodic capacity per subcarrier will be derived, while for the second scenario an approximation of the exact ergodic capacity value can be determined. Only the ergodic capacity of OFDM DF relay system implementing BTB SCP should be analyzed, as it is known that it is the scheme which maximizes the achievable capacity.

It is assumed that RS performs subcarrier ordering on both S-RS and RS-D links, all in accordance with subcarriers instantaneous SNRs, and then the mapping of appropriate subcarriers is applied, i.e. BTB SCP is realized. This means that the relation (7.3), for the PDF of SNR for the subcarrier with the k-th lowest SNR among the M total ones, can be applied for both links (hops). In the scenario with equal average SNRs on both hops, i.e. when $\bar{\gamma}_{SR} = \bar{\gamma}_{RD} = \bar{\gamma} = 1/\lambda$, the CDF of SNR for the k-th subcarrier on either of the two links, is given with:

$$F_k(x) = \sum_{i=0}^{k-1} \frac{\alpha_i}{\beta_i}(1 - e^{-\beta_i \lambda x}). \tag{7.20}$$

Substituting now the known PDF and CDF of SNR for the k-th subcarrier on both hops in the relation for the ordered statistics of random variables (3.2.1), the PDF of SNR for the k-th received subcarrier of the OFDM DF relay system with BTB SCP is obtained:

$$f_{DF,k}(x) = 2f_k^w(x)(1 - F_k(x))$$

$$= 2\lambda \left[\sum_{i=0}^{k-1} \alpha_i e^{-\beta_i \lambda x} - \sum_{i=0}^{k-1} \sum_{j=0}^{k-1} \frac{\alpha_i \alpha_j}{\beta_j}(e^{-\beta_i \lambda x} - e^{-(\beta_i+\beta_j)\lambda x}) \right]. \tag{7.21}$$

The ergodic capacity for the k-th subcarrier received at D in the considered scenario, $C_k = \mathbf{E}\{0, 5 \cdot \log_2(1 + x)\}$, may be evaluated using the following integral of the known solution, [81]:

$$\int_1^\infty e^{-\mu x} \ln x \, dx = \frac{1}{\mu} E_1(\mu), \quad \text{Re}\{\mu\} > 0, \tag{7.22}$$

which leads to:

$$C_k = \frac{1}{\ln(2)} \left[\sum_{i=0}^{k-1} \frac{\alpha_i E_1(\lambda\beta_i)}{\beta_i e^{\beta_i \lambda}} - \sum_{i=0}^{k-1} \sum_{j=0}^{k-1} \frac{\alpha_i \alpha_j}{\beta_j} \right.$$

$$\left. \times \left(\frac{E_1(\lambda\beta_i)}{\beta_i e^{\beta_i \lambda}} - \frac{E_1(\lambda(\beta_i + \beta_j))}{(\beta_i + \beta_j)e^{-(\beta_i+\beta_j)\lambda}} \right) \right]. \tag{7.23}$$

The total capacity of the OFDM DF relay system with BTB SCP is then obtained as a sum of capacities for all M subcarriers.

In the scenario with unequal average SNRs on hops, an approximation for the OFDM DF with BTB SCP system capacity may be used, being equal to half capacity of the link with the lower average SNR value. Factor 1/2 is taken here due to the fact that the communication process is realized in two time intervals. For the Rayleigh fading channel, using the integral given in (7.3), the ergodic capacity for the k-th subcarrier is then obtained as:

$$C_k = \frac{e^{\lambda_l}}{2\ln(2)} E_1(\lambda_l), \qquad (7.24)$$

where λ_l denotes the average SNR of the link which has the lower average SNR value.

7.5 Performance Analysis of OFDM DF Relay Systems with SCP

7.5.1 BER Performance Analysis of DPSK Modulated OFDM DF Relay Systems with SCP

BER performances of the DPSK modulated OFDM DF relay system with BTB SCP or BTW SCP, as well as BER performance of this system without SCP at the R station, are given in Figure 7.2. A scenario with the equal average SNRs on both hops is assumed. The complete matching of the analytically and simulation obtained BER graphs confirms the accuracy of the previously described analytical approach.

From Figure 7.2 it can be noticed that, for all values of the average SNRs on both hops, the OFDM DF relay system when BTW SCP is implemented can not achieve the level of BER attained when BTB SCP is applied. This means that for the SNR values of interest, the BTB SCP scheme should be used in OFDM DF relay systems for BER performance improvement. This is particularly important taking into account that the BTB SCP scheme also maximizes the achievable capacity in the considered relay system, which leads to the conclusion that the OFDM DF relay system with BTB SCP scheme presents an optimal solution from the point of BER performance improvement and capacity enhancement.

Performance comparison of the OFDM DF relay systems with BTB SCP and with no SCP shows that for the high SNR values BER graphs of these two systems slightly differ. However, it is significant to notice that the BTB

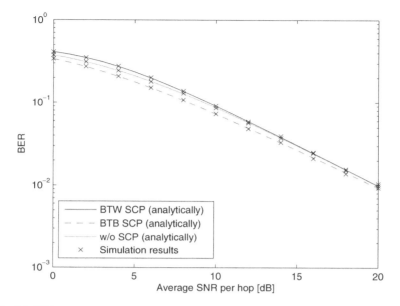

Fig. 7.2 BER performance of DPSK modulated OFDM DF relay system with and without SCP.

SCP scheme improves BER performances in the region of small SNR values on both hops, where it is most needed. The SNR gain of the considered relay system implementing BTB SCP is above 1 dB in the region of small SNRs on both hops, compared to the system without SCP.

7.5.2 BER Performance Analysis of BPSK Modulated OFDM DF Relay Systems with SCP

BER performances of the BPSK modulated OFDM DF relay system with implemented BTW SCP or BTB SCP, are shown in Figure 7.3. For the sake of comparison, BER graph for the OFDM DF relay system without SCP (w/o SCP) is also given. The scenario with the equal average SNRs on both hops is assumed. As it can be seen, all the analytically obtained results are verified through comparisons with the simulation results.

All the graphs given in Figure 7.3 approve the conclusion derived after the consideration of DPSK modulated OFDM DF relay systems with SCP, i.e. that the BTW SCP scheme should not be used in this type of relay systems, as it can not improve their BER performances. The level of BER performance

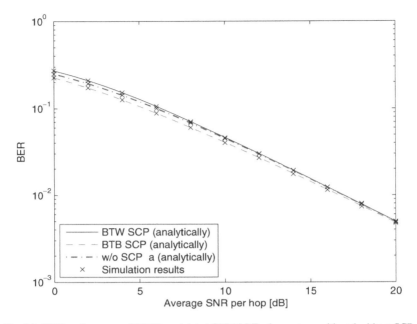

Fig. 7.3 BER performance of BPSK modulated OFDM DF relay system, with and without SCP.

improvement achieved through the BTB SCP incorporation in BPSK modulated OFDM DF relay systems is the highest in the region of small SNRs, and in spite of not having significant values, it is very important having in mind the fact that the same SCP scheme maximizes the system capacity.

7.5.3 Ergodic Capacity Analysis of OFDM DF Relay System with SCP

Figure 7.4 presents the average ergodic capacities for OFDM DF relay systems with and without (w/o) the BTB SCP scheme, in a scenario where the average values of SNRs on both hops are equal. The complete matching of the analytically derived results for OFDM DF relay system implementing BTB SCP with the simulation results confirms the validity of the conducted analytical approach. The average ergodic capacity per subcarrier is obtained through averaging ergodic capacity values over all $M = 16$ subcarriers of the considered system. In the case of OFDM DF relay systems without SCP, the average ergodic capacity is identical for all subcarriers, and it can be derived using the same approach as the one given in Chapter 7.4, while using appropriate

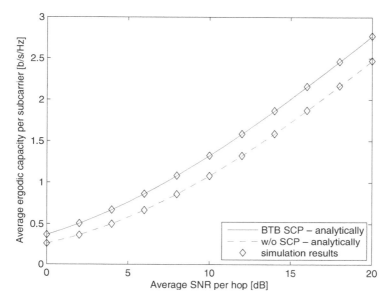

Fig. 7.4 Average ergodic capacity per subcarrier of OFDM DF relay system with and without SCP.

PDF of SNR and CDF of SNR functions that correspond to Rayleigh fading channel (relations (5.5) and (5.7)). With the described approach, the average ergodic capacity per subcarrier of OFDM DF relay systems without SCP can be obtained in the form:

$$C_k = \frac{e^{2\lambda}}{2\ln(2)} E_1(2\lambda). \tag{7.25}$$

Using Monte Carlo simulations, i.i.d. subcarriers on both hops are generated as complex Gaussian random variables with mean values equal to 0, and variances equal to 1/2, thus modeling Rayleigh fading on both hops, with the average subcarrier power being equal to 1. After the subcarrier ordering, done separately in the first hop and then on the second hop, all in accordance to their instantaneous SNRs, the subcarrier mapping is performed. For each mapped subcarrier pair, in each simulation realization, the subcarrier with the lower instantaneous SNR value is used for the ergodic capacity calculation. The presented average ergodic capacity graphs per subcarrier are obtained by averaging over 5000 simulation repetitions for each subcarrier pair, and then by averaging over $M = 16$ subcarrier pairs.

The comparison of the ergodic capacity graphs for OFDM DF relay systems with and without BTB SCP shows that significant capacity enhancement is achieved through implementation of BTB SCP scheme, as the SNR gain is about 2 dB for all the given values of the average ergodic capacity per subcarrier.

In the scenario with unequal average SNR values on hops, one half of the ergodic capacity of the channel with the lower average SNR may be used as an upper bound of ergodic capacity in the considered relay system. Through this approach a certain deviation of the exact ergodic capacity values appear in the cases when the average SNR values on both hops are approximately the same. This is confirmed with the graphs given in Figure 7.5, which represent average ergodic capacity values as a function of the average SNR on the RS-D link, for two different values of $\bar{\gamma}_{SR}$. The simulation obtained graphs, as well as the graphs representing the upper bound obtained analytically using the relation (7.24), are shown. It is clear that very tight upper bounds are derived, which differ from the real ergodic capacity values only in the case when $\bar{\gamma}_{SR} \approx \bar{\gamma}_{RD}$.

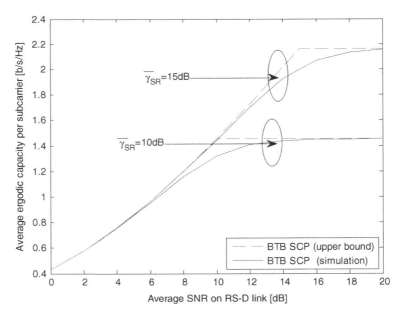

Fig. 7.5 Average ergodic capacity per subcarrier for OFDM DF relay system with SCP.

7.6   Comparative Performance Analysis of OFDM AF and OFDM DF Relay Systems with SCP

7.6.1   Comparative BER Performance Analysis of OFDM AF and OFDM DF Relay Systems with SCP

Figure 7.6 shows BER graphs for DPSK modulated OFDM AF FG relay system with SCP and OFDM DF relay system with BTB SCP, in the case when average SNR on the S-RS link is equal to the average SNR on the RS-D link. As the OFDM AF VG relay system implementing BTB SCP achieves better BER performance than the OFDM AF FG relay system with SCP in the region of small SNRs, BER performance of this system is also given. From the presented graphs it is obvious that the OFDM DF relay system with BTB SCP achieves the lowest BER values for SNR values on both hops below 10 dB, while for the higher SNR values the OFDM AF FG relay system implementing BTW SCP has the best BER performance. In the region of higher SNR values on both hops, the OFDM AF FG relay system with BTW SCP achieves significantly better performance than all the other analyzed systems. Thus, for example, for

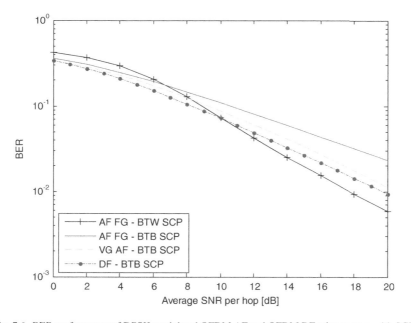

Fig. 7.6   BER performance of DPSK modulated OFDM AF and OFDM DF relay system with SCP.

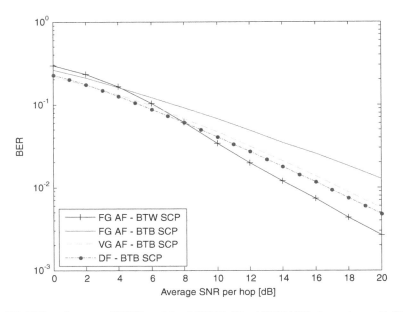

Fig. 7.7   BER performance of BPSK modulated OFDM AF and OFDM DF relay systems with SCP.

BER value of 10^{-2}, this system attains the SNR gain of above 2 dB comparing to the OFDM DF relay system with BTB SCP.

The same conclusions may be derived if BER performances of the BPSK modulated OFDM relay systems with SCP are considered (Figure 7.7). The only difference is that the boundary SNR values above which OFDM AF FG relay systems outperforms all the other systems is here even lower (below 8 dB).

In mobile cellular systems of the next generation, infrastructure based relay stations will be deployed, and they will be placed in a way to ensure slow variations of the channel conditions on backhaul link (base station-to-R station). Thus, it is interesting to analyze BER performance of the considered systems as a function of the average SNR on the RS-D link, which in downlink communication process corresponds to communication between the R station and a mobile terminal (user). BER graphs given in Figures 7.8 and 7.9 show that for the average SNR on the S-RS link being equal to 15 dB, the OFDM AF FG relay system with BTB SCP represents an optimal solution in terms of the achieved BER values for all SNR values on the RS-D link. However, when the average SNR on the S-RS link is low (for example 5 dB on the given

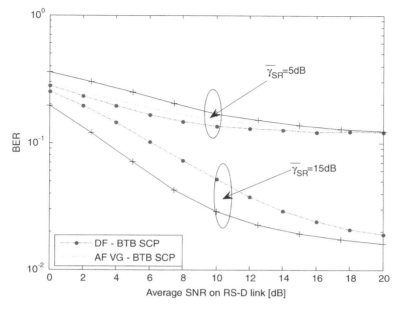

Fig. 7.8 BER performance of DPSK modulated OFDM AF and OFDM DF relay systems with SCP ($\bar{\gamma}_{SR} = 5$ dB and $\bar{\gamma}_{SR} = 15$ dB).

figures), the OFDM DF relay system with BTB SCP achieves the best BER performance, regardless the SNR values on the RS-D link.

Based on the BER graphs given in the above figures, it can be concluded that, when BER performance is used as the main selection criteria, the OFDM DF relay system with BTB SCP can be considered as an optimal solution for the configuration of the Type II relay stations in the next generation WWAN networks, in the region of small average SNR values on the S-RS link (approximately up to 10 dB in the case of DPSK modulation). However, for the higher SNR values on the S-RS link it is the OFDM AF FG relay system with BTW SCP that provides better solution. The SNR value where one system outperforms another varies, depending on the type of modulation applied.

7.6.2 Comparative Analysis of Ergodic Capacities of OFDM AF and OFDM DF Relay Systems with SCP

The system capacity is one of the most important performance metrics for wireless communication systems. Therefore, it is necessary to compare

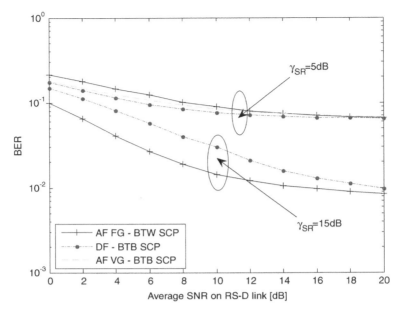

Fig. 7.9 BER performance of BPSK modulated OFDM AF and OFDM DF relay systems with SCP ($\bar{\gamma}_{SR} = 5$ dB and $\bar{\gamma}_{SR} = 15$ dB).

different OFDM based relay systems with SCP from the point of achievable capacity. Figure 7.10 presents the graphs of the average ergodic capacity per subcarrier for OFDM AF FG, OFDM AF VG and OFDM DF relay systems with the BTB SCP scheme, in the scenario with equal average SNRs on both hops. All the presented graphs are obtained through simulations. From the figure, it can be seen that the OFDM DF relay system with BTB SCP achieves the highest capacity, in the scenario considered. For the average SNR per hop values up to 2,5 dB, the OFDM AF FG relay system has the lowest capacity, while for the higher average SNR per hop values the OFDM AF VG has the worst performance. The difference in capacity between the OFDM DF and OFDM AF FG relay systems decreases with an increase of the average SNR per hop. Thus, for example, for the average ergodic capacity per subcarrier of 0,5 b/s/Hz, the SNR gain of the OFDM DF relay system with BTB SCP is around 2 dB when compared with the OFDM AF FG relay system with BTB SCP, while for the average ergodic capacity of 2 b/s/Hz this SNR gain is less than 1 dB.

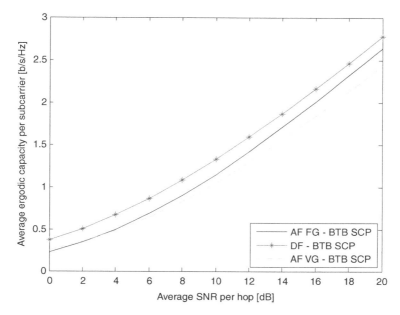

Fig. 7.10 Average ergodic capacity per subcarrier for OFDM AF and OFDM DF relay systems.

Comparison of the average ergodic capacities per subcarrier, presented as a function of the average SNR on the RS-D link, is given in Figure 7.11. It can be noticed that for the average SNR values on the S-RS link of up to 5 dB, the OFDM DF relay system with BTB SCP achieves the highest capacity, irrespective of the SNR value on the RS-D link. This should be emphasized having in mind the earlier shown fact that the OFDM DF relay system with BTB SCP has the best BER performance in this region of SNR values. For the higher SNR values on the S-RS link, a SNR region can be recognized where the OFDM AF FG relay system with BTB SCP achieves the highest capacity. For example, when $\bar{\gamma}_{SR} = 15$ dB this region is defined with the SNR values on the RS-D link being between 0 dB and 12 dB. The advantage of the OFDM AF FG relay system is particularly high for the small values of $\bar{\gamma}_{RD}$, thus having the capacity of this system to be two times higher than the capacity of the OFDM DF relay system with BTB SCP, when $\bar{\gamma}_{RD} = 0$ dB and $\bar{\gamma}_{SR} = 15$ dB.

Based on the presented graphs of average ergodic capacity per subcarrier for the considered OFDM relay systems with BTB SCP, it can be concluded that in order to achieve the highest capacity the OFDM DF relay system should

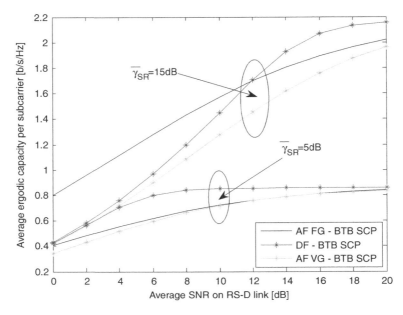

Fig. 7.11 Average ergodic capacity per subcarrier for OFDM AF and OFDM DF relay systems ($\bar{\gamma}_{SR} = 5$ dB and $\bar{\gamma}_{SR} = 15$ dB).

Table 7.1. $\bar{\gamma}_{RD}$ regions where OFDM AF FG relay systems with BTB SCP achieves the highest capacity, for a given $\bar{\gamma}_{SR}$.

$\bar{\gamma}_{SR}$ [dB]	6	8	10	12	14	16	18	20
$\bar{\gamma}_{RD}$ [dB]	$\leq 0{,}5$	$\leq 3{,}5$	≤ 6	$\leq 8{,}5$	≤ 11	≤ 13	≤ 15	$\leq 17{,}5$

be implemented. The OFDM AF VG relay system should not be considered at all as an option, while in certain conditions the OFDM AF FG relay system may outperform the OFDM DF relay system in terms of achievable capacity. The region of average SNR on the RS-D link, for the given $\bar{\gamma}_{SR}$, where the OFDM AF FG relay system with BTB SCP achieves the highest capacity is given in Table 7.1.

From Table 7.1 it can be seen that starting from $\bar{\gamma}_{SR} = 6$ dB the OFDM AF FG relay system with BTB SCP outperforms in terms of achievable capacity the OFDM DF relay system implementing BTB SCP. With an increase of the average SNR on the S-RS link, the SNR region where the OFDM AF FG relay system presents an optimal solution also increases. Thus, for example, for $\bar{\gamma}_{SR} = 10$ dB this region has the upper value of 6 dB, while for $\bar{\gamma}_{SR} = 20$ dB the OFDM AF FG relay system with BTB SCP outperforms all the other considered systems in the region of $\bar{\gamma}_{RD}$ values up to 17,5 dB.

Based on the given comparative BER and ergodic capacity performances analyses for all presented OFDM relay systems with SCP, some general conclusions can be drawn when optimal solutions for the Type II relay stations in the next generation WWAN networks are considered. Assuming that the average SNR on the S-RS link is less than approximately 8 dB, then the OFDM DF relay system with BTB SCP can be considered as an optimal relay station configuration, as it achieves the lowest BER and the highest capacity values. For $\bar{\gamma}_{SR} \geq 8$ dB, some regions where this system still presents an optimal solution can be identified, and their boundaries can be seen from Table 7.1.

For all other values of the average SNR on the S-RS and RS-D links, the OFDM AF FG relay system with an appropriate SCP scheme can be recognized as an optimal solution for the Type II relay stations. Thus, if the capacity is the performance of interest, then the BTB SCP scheme should be chosen. If BER is the performance of interest, then for the higher $\bar{\gamma}_{SR}$ values the BTW SCP scheme should be implemented. There, as it is shown in subchapters 5.5.1 and 5.5.2, the lower $\bar{\gamma}_{SR}$ boundary depends on the type of the modulation scheme applied and, for example, for BPSK it is 10 dB.

References

[1] M. Uysal and M. M. Fareed, "Cooperative Diversity Systems for Wireless Communication," Chapter in Handbook on Information and Coding Theory, I. Woungang, S. Misra and S. C. Misra (Eds.), World Scientific, March 2010.

[2] Z. Zhou, S. Zhou, J. Cui, and S. Cui, "Energy-efficient cooperative communication based on power control and selective single-relay in wireless sensor networks," *IEEE Transactions on Wireless Communications*, vol. 7, no. 8, 3066–3078, 2008.

[3] M. Dohler and J. Li, *Cooperative Communications: Hardware, Channel & PHY*, Wiley, 2010.

[4] T. Imich, D. Schultz, R. Pabst, and P. Wienert, "Capacity of a Relaying Infrastructure for Broadband Radio Coverage of Urban Areas," in *Proceedings of 10th WWRF Meeting*, New York. http://www.comnets.rwth-aachen.de/, Oct 2003.

[5] N. Esseling, E. Weiss, A. Kraemling, and W. Zinvas, "A Multi Hop Concept for Hiper-LAN/2: Capacity and Interference," *in Proceedings of European Wireless 2002*, vol. 1, pp. 1–7, Florence, Italy, Feb. 2002.

[6] S. Yatawatta, A. P. Petropulu, "A Multiuser OFDM System with User Cooperation," in *Proceedings of 38th Asilomar Conference on Signals, Systems, and Computers*, pp. 319–323 Vol. 1, 2004.

[7] A. Nosratinia and A. Hedayat, "Cooperative Communications in Wireless Networks," *IEEE Comm. Magazine*, 74–80, October 2004.

[8] E. C. van der Meulen, *Transmission of Information in a T-terminal Discrete Memoryless Channel*, PhD thesis, Dept. of Statistics, University of California, Berkeley, 1968.

[9] E. C. van der Meulen, "Three-Terminal Communication Channels," *Advanced Applied Probability*, vol. 3, pp. 120–154, 1971.

[10] N. Esseling, B.H. Walke, and R. Pabst, "Performance Evaluation of a Fixed Relay Concept for Next Generation Wireless Systems," in *Proceeding of 15th IEEE International Symposium PIMRC 2004*, Vol. 2, pp. 744–751.

[11] K. Doppler, "Integration of Amplify and Forward Relays in an OFDM Network," in *Proceedings of 39th Asilomar Conference on Signals, Systems, and Computers*, pp. 1471–1475, Pacific Grove, 2005.

[12] G. Li and J. Liu, "On the Capacity of the Broadband Relay Networks," in *Proceedings of 38th Annual Asilomar Conference on Signals, Systems and Computers*, CA, USA, November 2004.

[13] F. Ng, X. Li, "Cooperative STBC-OFDM Transmissions with Imperfect Synchronizations in Time and Frequency," in *Proceedings of 39th Asilomar Conference on Signals, Systems, and Computers*, Pacific Grove, 2005.

[14] K. Doppler, A. Hottinen, "Multi-Antenna Relay Nodes in OFDM Systems," in *Proceedings of ISWCS 2006*, pp. 258–261.

[15] G. Li, H. Liu, "Resource Allocation for OFDMA Relay Networks," in *Proceedings of 38th Asilomar Conference on Signals, Systems, and Computers*, pp. 203–1207 Vol. 1, 2004.

[16] G. Li, H. Liu, "Resource Allocation for OFDMA Relay Networks with Fairness Constraints," *IEEE Journal on Selected Areas in Communications*, Volume. 24, Issue. 1, pp. 2061–2069.

[17] Y. Guan-Ding, Z. Zhao-Jang, C. Yan, C. Shi, and Q. Pei-Liang, "Power Allocation for Non-Regenerative OFDM Relaying Channels," in *Proceedings of WCNMC 2005 Conference*, vol. 1, pp. 185–188, 2005.

[18] B. Gui, L.J. Cimini, and L. Dai, "OFDM for Cooperative Networking with Limited Channel State Information," in *Proceedings of MILCOM 2006 Conference*, pp. 1–6, Washington, Oct. 2006.

[19] I. Hammerstrom and A. Wittneben, "On the Optimal Power Allocation for Nonregenerative OFDM Relay Links," in *Proceedings of ICC 2006 Conference*, pp. 4463–4468, Istanbul, June 2006.

[20] A. Hottinen and T. Heikkinen, "Subchannel Assignment in OFDM Relay Nodes," *in Proceedings of 40th Annual Conference on Information Sciences and Systems*, 2006.

[21] I. Hammerstrom and A. Wittneb, "Joint Power Allocation for Non-Regenerative MIMO-OFDM Relay Links," in *Proceedings of IEEE International Conference on Acoustic, Speech and Signal Processing*, May 2006.

[22] M. Herdin, "A Chunk Based OFDM Amplify-and-Forward Relaying Scheme for 4G Mobile Radio Systems," in *Proceedings of IEEE International Conference on Communications* (ICC 2006), Istanbul, Turkey, 2006.

[23] A. Hottinen and T. Heikkinen, "Optimal Subchannel Assignment in a Two-Hop OFDM Relay," in *Proceedings of IEEE 8th Workshop on Signal Processing Advances in Wireless Communications*, 2007.

[24] M. Herdin and G. Auer, "Pilot Design for OFDM Amplify-and-Forward with Chunk Reordering," in *Proceedings of WCNC 2007 Conference*, pp. 1401–1406.

[25] C. K. Ho and A. Pandharipande, "BER Minimization in Relay-Assisted OFDM Systems by Subcarrier Permutation," in *Proceedings of IEEE Vehicular Technology Conference* (VTC 2008), Singapore, 2008.

[26] C. R. N. Athaudage, M. Saito, and J. Evans, "Performance Analysis of Dual-Hop OFDM Relay Systems with Subcarrier Mapping," in *Proceedings of IEEE International Conference on Communications* (ICC 2008), Beijing, China, 2008.

[27] T. Riihonen, R. Wichman, J. Hamalainen, and A. Hottinen, "Analysis of Subcarrier Pairing in a Cellular OFDMA Relay Link," in *Proceedings of ITG Workshop on Smart Antennas*, WSA 08, pp. 104–111, Viena, 2008.

[28] E. Kocan, M. Pejanovic-Djurisic, D. S. Michalopoulos, and G. K. Karagiannidis, "BER Performance of OFDM Amplify-and-Forward Relay System with Subcarrier

Permutation," in *Proceedings of the IEEE Wireless VITAE 2009 Conference*, pp. 252–256, Aalborg, Denmark, May 2009.

[29] E. Kocan, M. Pejanovic-Djurisic, D. S. Michalopoulos, and G. K. Karagiannidis, "Performance Evaluation of OFDM Amplify-and-Forward Relay System with Subcarrier Permutation," *IEICE Trans. on Communications*, Vol. E93-B, No. 05, pp. 1216–1223, May. 2010.

[30] E. Kocan, M Pejanovic-Djurisic, and Z.Veljovic, "BER Performance of M-QAM Modulated OFDM Based Relay System with Subcarrier Mapping," in *proceedings of WPMC 2010 Conference*, Recife, Brazil, Oct. 2010.

[31] E. Kocan and M Pejanovic-Djurisic, "OFDM AF variable gain relay system for the next generation mobile cellular networks", *paper accepted for publication in TELFOR Journal*, vol. 4, 2012.

[32] E. Kocan, M Pejanovic-Djurisic, "Performance Enhancement of OFDM Non-Regenerative Variable Gain Relay Systems," in *Proceedings of IEEE conf. MELECON 2012*, pp. 1038–1041,Yasmine Hammamet, Tunisia, March 2012.

[33] E. Kocan, M Pejanovic-Djurisic, Z. Veljovic, "On the Optimal Subcarrier Mapping Scheme in OFDM Decode-and-Forward Relay Systems," In *Proceedings of IEEE WTS 2011*, pp. 1–5, New York, USA, April 2011.

[34] W. Ying, Q. Xin-chun, W. Tong, and L. Bao-Ling, "Power Allocation and Subcarrier Pairing Algorithm for Regenerative OFDM Relay Systems," in *Proceedings of VTC 2007 — Spring Conference*, pp. 2727–2731, Dublin, April 2007.

[35] B. Gui and L.J. Cimini Jr., "Bit Loading Algorithms for Cooperative OFDM Systems," in *Proceedings of MILCOM 2007 Conference*, pp. 1–7.

[36] D. Chen and J. N. Laneman, "Joint Power and Bandwidth Allocation in Multihop Wireless Networks," in *Proceedings of WCNC 2008 Conference*, pp. 990–995.

[37] W. Wang, S. Yang, L. Gao, "Comparison of Schemes for Joint Subcarrier Matching and Power Allocation in OFDM Decode-and-Forward Relay Systems," in *Proceedings of ICC 2008*, pp. 4983–4987, Beijing.

[38] C. K. Ho, R. Zhang, and Y-C Liang, "Two-Way Relaying over OFDM: Optimized Tone Permutation and Power Allocation," in *Proceedings of ICC 2008*, Beijing.

[39] Y. Li, W. Wang, J. Kong, W. Hong, X. Zhang, and M. Peng, "Power Allocation and Subcarrier Pairing in OFDM Based Relaying Networks," in *Proceedings of ICC 2008*, Beijing.

[40] B. Can, H. Yomo, and E. de Carvalho, "Hybrid Forwarding Scheme for Cooperative Relaying in OFDM Based Networks," in *Proceedings of ICC 06 Conference*, pp. 4520–4525, Istanbul, 2006.

[41] Basak Can *et al.*, "Hybrid Forwarding Apparatus and Method for Cooperative Relaying in an OFDM Network," Korean Intellectual Property Office. Patent No.: KR2006-0033844. Apr. 14, 2006.

[42] L. Dai, B. Gui, and L.J. Cimini, "Selective Relaying in OFDM Multihop Cooperative Networks," in *Proceedings of IEEE conference WCNC 2007*, pp. 964–969.

[43] L. Dai, B. Gui, and L.J. Cimini, "Selective Relaying in Cooperative OFDM Systems: Two-Hop Random Nework," in *Proceedings of IEEE conference WCNC 2008*, pp. 996–1001.

[44] H. Suraweera and J. Armstrong, "Performance of OFDM-based Dual-Hop Amplify-and-Forward Relaying," *IEEE Communication Letters*, vol. 11, no. 9, 726–728, Sept. 2007.

[45] M. Saito, C. N. Authage, and J. Evans, "On Power Allocation for Dual-Hop Amplify-and-Forward OFDM Relay Systems," in *Proceedings of GLOVECOM 2008 Conference*, pp. 1–6, New Orleans, 2008.

[46] Y. Ma, N. Yi, and R. Tafazolli, "Bit and Power Loading for OFDM-Based Three-Node Relaying Communications," *IEEE Transaction on Signal Processing*, Vol. 56, Issue 7, 3236–3247, July 2008.

[47] M. Ibrahimi and B. Liang, "Efficient Power Allocation in Cooperative OFDM System with Channel Variation," in *Proceedings of ICC 2008 Conference*, pp. 3022–3028, Beijing.

[48] L. Vandendorpe, R. T. Duran, J. Louveaux, and A. Zaidi, "Power Allocation for OFDM Transmission with DF Relaying," in *Proceedings of IEEE Symposium on Communications and Vehicular Technology in the Benelux*, pp. 1–6, Delft, 2007.

[49] Y. Yang, H. Hu, J. Xu, and G. Mao, "Relay Technologies for WiMAX and LTE-Advanced Mobile Systems," *IEEE Communication Magazine*, vol. 10, 100–105, October 2009.

[50] S. W. Peters, A. Y. Panah, K. T. Truong, and R. W. Heath Jr., "Relay Architectures for 3GPP LTE-Advanced," *EURASIP Journal on Wireless Communications and Networking*, Volume 2009, Article ID 618787.

[51] S. W. Peters and R. W. Heath, "The Future of WiMAX: Multihop Relaying with IEEE 802.16j," *IEEE Communications Magazine*, vol. 1, 104–111, January 2009.

[52] C.-X. Wang, X. Hong, X. Ge, X. Cheng, G. Zhang, and J. Thompson, "Cooperative MIMO Channel Models: A survey," *IEEE Communications Magazine*, vol. 2, 80–87, February 2010.

[53] D. Soldani and S. Dixit, "Wireless Relays for Broadband Access," *IEEE Communications Magazine*, vol. 3, 58–66, March 2008.

[54] J. Sydir and R. Taori, "An Evolved Cellular System Architecture Incorporating Relay Stations," *IEEE Communications Magazine*, 115–121, June 2009.

[55] T. Saito, Y. Tanaka, T. Kata, "Trends in LTE-WiMAX Systems," *FUJITSU Scient. Tech. Journal*, Vol. 45, No. 4, 352–362, October 2009.

[56] D. Niyato, E. Hossain, D. I. Kim, "Relay-Centric Radio Resource Management and Network Planning in IEEE 802.16j Mobile Multihop Relay Networks," *IEEE Transactions on Wireless Communications*, Vol. 8, No. 12, 6115–6125, December 2009.

[57] B. Can, H. Yanikomeroglu, F. A. Onat, E. De Carvalho, and H. Yomo, "Efficient Cooperative Diversity Schemes and Radio Resource Allocation for IEEE 802.16j," in *Proceedings of WCNC 2008*, pp. 36–41, Las Vegas, April 2008.

[58] IEEE Standard for Information Technology–Telecommunications and information exchange between systems–Local and metropolitan area networks–Specific requirements Part 11: Wireless LAN Medium Access Control (MAC) and Physical Layer (PHY) specifications Amendment 10: Mesh Networking. *Institute of Electrical and Electronics Engineers* / 10-Sep-2011.

[59] *Cooperation in Wireless Networks*, edited by Frank H. P. Fitzek, Marcos D. Katz, Springer 2006.

[60] T. M. Cover, and A. A. El Gamal, "Capacity Theorems for the Relay Channel," *IEEE Transactions on Information Theory*, vol. 25, 572–584, 1979.

[61] J. N. Laneman, D. N. C. Tse, and G. W. Wornell, "Cooperative Diversity in Wireless Networks: Effcient Protocols and Outage Behavior," *IEEE Transactions on Information Theory*, vol. 50, 3062–3080, December 2004.

[62] M. O. Hasna and M. S. Alouini, "A Performance Study of Dual-Hop Transmissions with Fixed Gain Relays," *IEEE Transactions on Wireless Communications*, vol. 3, 1963–1968, November 2004.

[63] G. K. Karagiannidis, "Performance Bounds of Multihop Wireless Communications with Blind Relays Over Generalized Fading Channels," *IEEE Transactions on. Wireless Communications*, vol. 5, 498–503, March 2006.

[64] H. Shin, and J B. Song, "MRC Analysis of Cooperative Diversity with Fixed-Gain Relays in Nakagami-m Fading Channels," *IEEE Transactions on Wireless Communications*, vol. 7, no. 6, 2069–2074.

[65] G. Farhadi, and N. C. Beaulieu, "On the Outage and Error Probability of Amplify-and-Forward Multi-Hop Diversity Transmission Systems," in *Proceedings of ICC'08 Conference*, Beijing, 2008.

[66] J. N. Laneman, *Cooperative Diversity in Wireless Networks: Algorithms and Architectures*, PhD dissertation, Massachusetts Institute of Technology 2002.

[67] R. U. Nabar, H. Bolcskei, and F. W. Kneubuhler, "Fading Relay Channels: Performance Limits and Space-Time Signal Design," *IEEE Journal on Selected Areas in Communications*, pp. 1099–1109, 2004.

[68] C. S. Patel and G. L. Stuber, "Channel Estimation for Amplify and Forward Relay Based Cooperation Diversity Systems," *IEEE Transactions on Wireless Communications*, vol. 6, no. 6, 2348–2356, June 2007.

[69] C. S. Patel, G. L. Stuber, and T. G. Pratt, "Statistical Properties of Amplify and Forward Relay Fading Channels," *IEEE Transactions on Vehicular Technology*, vol. 55, no. 1, 1–9, January 2006.

[70] G. Farhadi and N. C. Beaulieu, "On the Ergodic Capacity of Wireless Relaying Systems over Rayleigh Fading Channels," *IEEE Transactions on Wireless Communications*, vol. 7, no. 11, 4462–4467, Nov. 2008.

[71] A. Ribeiro, X. Cai, and G. B. Giannakis, "Symbol Error Probability for General Cooperative Links," *IEEE Transactions on Wireless Communications*, vol. 4, no. 3, 1264–1273, May 2005.

[72] M. O. Hasna and M. S. Alouini, "End-to-End Performance of Transmission Systems with Relays Over Rayleigh-Fading Channels," *IEEE Transactions on Wireless Communications*, vol. 2, no. 6, 1126–1131, Nov. 2003.

[73] G. K. Karagiannidis, T. A. Tsiftsis, and R. K. Malik, "Bounds for Multihop Relayed Communications in Nakagami-m Fading," *IEEE Transactions on Communications*, vol. 54, no. 1, 18–22, Jan. 2006.

[74] D. S. Michalopoulos and G. K. Karagiannidis, "PHY-Layer Fairness in Amplify and Forward Cooperative Diversity Systems," *IEEE Transactions on Wireless Communications*, vol. 7, no. 3, 1073–1083, March 2008.

[75] H. A. Suraweera, R. Louie, Y. Li, G. K. Karagiannidis, and B. Vucetic, "Two Hop Amplify-and-Forward Transmission in Mixed Rayleigh and Rician Fading Channels," *IEEE Transactions on Communications*, Vol. 13, no. 4, April 2009.

[76] T. Wang, A. Cano, G. B. Giannakis, and J. N. Laneman, "High-Performance Cooperative Demodulation with Decode-and-Forward Relays," *IEEE Transactions on Communications*, vol. 55, no. 7, July 2007.

[77] K. Azarian, H. El Gamal, and P. Schniter, "On the Achievable Diversity-Multiplexing Tradeoff in Half-Duplex Cooperative Channels," *IEEE Transactions on Information Theory*, vol. 51, no. 12, 4152–4172, December 2005.

[78] P. Mitran, H. Ochia, and V. Tarokh, "Space-Time Diversity Enhancements using Cooperative Communications," *IEEE Transactions on Information Theory*, vol. 51, no. 6, 2041–2057, June 2005.

[79] M. Abramovitz and I. A. Stegun, *Handbook of Mathematical Functions with Formulas, Graphs, and Mathematical Tables*, 9th ed. New York: Dover, 1972.

[80] A. Papoulis, *Probability, Random Variables, and Stochastic Processes*, 3rd ed. McGraw-Hill, 1991.

[81] I. S. Gradshteyn and I. M. Ryzhik, *Table of Integrals, Series, and Products*, 6th ed. New York: Academic, 2000.

[82] M. K. Simon and M.-S. Alouini, *Digital Communication over Fading Channels*, 2nd ed. New York: Wiley, 2005.

[83] H. Sato, *Information Transmission through a Channel with Relay*, The Aloha System, University of Hawaii, Honolulu, Technical Report B76-7, March 1976.

[84] A. Sendonaris, E. Erkip, and B. Aazhang, "Increasing Uplink Capacity via User Cooperation Diversity," in *Proceedings of IEEE International Symposium on Information Theory*, Aug. 1998.

[85] A. Sendonaris, E. Erkip, B. Aazhang, "User Cooperation Diversity. Part I. System Description," *IEEE Transactions on Communications*, vol. 51, 1927–1938, 2003.

[86] A. Sendonaris, E. Erkip, and B. Aazhang, "User Cooperation Diversity. Part II. Implementation Aspects and Performance Analysis," *IEEE Transactions on Communications*, vol. 51, 1939–1948, 2003.

[87] J. N. Laneman, G. W. Wornell, "Distributed Space-Time Coded Protocols for Exploiting Cooperative Diversity in Wireless Networks," *IEEE Transactions on Information Theory*, vol. 49, no. 10, 2415–2425, 2003.

[88] M. A. Khojastepour, *Distributed Cooperative Communications in Wireless Networks*, PhD thesis, Dept. of Electrical and Computer Engineering, Rice University, 2004.

[89] A. Stefanov, E. Erkip, "Cooperative Coding for Wireless Networks," *IEEE Transactions on Communications*, vol. 52, 1470–1476, 2004.

[90] R. Prasad, *OFDM for Wireless Communications Systems*, 1st ed. Boston, MA: Artech House Inc., 2004.

[91] P. H. Moose, "A Technique for Orthogonal Frequency Division Multiplexing Frequency Offset Correction," *IEEE Transactions on Communications*, vol. 42, No. 10, October 1994.

[92] E. Kocan, M. Pejanovic-Djurisic, "Channel Estimation Technique with Frequency Offset Correction for OFDM Space-Time Diversity System," *Proceeding of the TELSIKS Conference*, vol. 1, 93–96, Nis, September 2005.

[93] E. Kocan, M. Pejanovic-Djurisic, M. Ilic, "A Novel Frequency Synchronization Method for OFDM System with Frequency Domain Selection Combining Diversity," in *Proceeding of the ISWCS Conference*, pp. 796–799, Valencia, September 2006.

[94] E. Kocan, M Pejanovic-Djurisic, and Z. Veljovic, "Efficient Frequency Synchronization and Channel Estimation Method for OFDM Wireless Systems," in *Proceedings of MELECON 2010 Conference*, pp. 487–491, Valletta, Malta, April 2010.

[95] S. H. Han, J. L. Lee, "An Overview of Peak-to-Average Power Ratio Reduction Techniques for Multicarrier Transmissions," *IEEE Wireless Communications*, vol. 12, no. 2, pp. 56–65, April 2005.

[96] G. R. Hiertz, D. Denteneer, L. Stibor, Y. Zang, X. P. Costa, and B. Walke, "The IEEE 802.11 Universe," *IEEE Communications Magazine*, pp. 62–10, January 2010.

[97] IEEE Standard for Local and Metropolitan Area Networks, Part 16: Air Interface for Broadband Wireless Access Systems, IEEE Std 802.16-2009, 29. May 2009.

[98] http://www.mobilecomms-technology.com/projects/t-mobile/

[99] S. Galli and O. Logvinov, "Recent Developments in Standardization of Power Line Communications within IEEE," *IEEE Communications Magazine*, pp. 64–71, July 2008.

[100] http://www.3gpp.org/Release-8

[101] Q. Li, G. Li, W. Lee, M. Lee, D. Mazzarese, B. Clerckx, Z. Li, "MIMO Techniques in WiMAX and LTE: A Feature Overview," *IEEE Communications Magazine*, 86–91, May 2010.

[102] J. G. Proakis, *Digital Communications*, McGraw Hill Higher Education, 4th Edition, 2000.

[103] P. Moberg *et al.*, "Performance and Cost Evaluation of Fixed Relay Nodes in Future Wide Area Cellular Networks," *IEEE PIMRC 2007*, Athens, Sept. 2007.

[104] C. Hozmann, M. Dittrich, and S. Goebbels, "Dimensioning and Capacity Evaluation of Cellular Multihop WiMAX Networks," in *Proceedings of IEEE Mobile WiMAX*, March 2007.

[105] IEEE Standard for Local and Metropolitan Area Networks, Part 16: Air Interface for Broadband Wireless Access Systems; Amendment 1: Multihop Relay Specification, IEEE Std 802.16j, 12. June 2009.

[106] J. Sydir at all, "Harmonized Contibution on 802.16j Usage Models," *IEEE C802.16j-06/015*, 2006.

[107] K. J. Miyahara, "IMT-Advanced: Objectives and Challenges," in *Proceedings of CISC'09*, pp.1-4, Greensboro, Oct. 2009.

[108] M. Baker (Alcatel-Lucent), "LTE-Advanced Physical Layer, Rev. 090003r1," *IMT-Advanced Evaluation Workshop*, 17–19 December 2009, Bejing.

[109] IEEE Standard for local and metropolitan area networks — Part 16:Air Interface for Broadband Wireless Access Systems. Amendment 3: Advanced Air Interface. May 2011.

[110] S. Ahmadi, "An Overview of Next Generation Mobile WiMAX Technology," *IEEE Communications Magazine*, 84–98, June 2009.

[111] M. Chiani, D. Dardari, and M. K. Simon, "New exponential bounds and approximations for the computation of error probability in fading channels," *IEEE Transactions on Wireless Communications*, vol. 2, 840–845, July 2003.

[112] K. Loa, C.-C. Wu *et al.*, "IMT-Advanced Relay Standards," *IEEE Communication Magazine*, 40–48, Aug. 2010.

Index

RIVER PUBLISHERS SERIES IN COMMUNICATIONS

Lightning Source UK Ltd.
Milton Keynes UK
UKOW07f1034171214

243278UK00001B/10/P

9 788792 329271